ROBOTIC TECHNOLOGY

ROBOTIC TECHNOLOGY

Edited by
A. Pugh

Published by Peter Peregrinus Ltd., London, UK.

©1983: Peter Peregrinus Ltd

While the author and the publishers believe that the information and
guidance given in this work is correct, all parties must rely upon their own
skill and judgment when making use of it. Neither the author nor the
publishers assume any liability to anyone for any loss or damage caused by
any error or omission in the work, whether such error or omission is the
result of negligence or any other cause. Any and all such liability is
disclaimed.

British Library Cataloguing in Publication Data

Robotic technology.
  1. Robots, Industrial
  I. Pugh, A.
  629.8'92   TS191

ISBN 0—86341—004—9

Printed in England by Short Run Press Ltd., Exeter

Contents

List of contributors

Dr E A Appleton
University Engineering Department
Trumpington Street
Cambridge
CB2 1PZ

Mr P G Davey
SERC Robotics Co-ordinator
Interactive Computing Facility
Rutherford Laboratory
Chilton
Didcot
Oxfordshire
OX11 0QX

Mr B L Davies
Department of Mechanical Engineering
University College
Gower Street
London
WC1E 6BT

Mr M Dooner
Department of Production Engineering
Lanchester Polytechnic
Priory Street
Coventry
CV1 5FB

Dr P J Drazan
Department of Mechanical Engineering
   and Engineering Production
UWIST
King Edward VII Avenue
Cardiff
CF1 3NU

Dr M Edkins
Lecturer in Mechanical Engineering
Department of Mechanical Engineering
Manufacturing and Machine Tools
   Division
UMIST
PO Box 88
Sackville Street
Manchester
M60 1QD

Dr J R Hewit
Department of Mechanical Engineering
Stephenson Building
Newcastle upon Tyne
NE1 7RU

Dr M H E Larcombe
Senior Lecturer
Department of Computer Science
University of Warwick
Coventry
CV4 7AL

Miss E McLellan
Senior Research Officer
School of Production Studies
Cranfield Institute of Technology
Cranfield
Bedford
MK43 0AL

Dr C J Page
Senior Lecturer
Department of Production Engineering
Lanchester Polytechnic
Priory Street
Coventry
CV1 5FB

Professor A Pugh
Head
Department of Electronic Engineering
University of Hull
Cottingham Road
Hull
HU6 7RX

Dr R J Poppleston & Miss A P Ambler
Department of Artificial Intelligence
University of Edinburgh
Forrest Hill
Edinburgh
EH1 2QL

Dr. G T Russell
Senior Lecturer
Department of Electrical and
    Electronic Engineering
Heriot-Watt University
31–35 Grassmarket
Edinburgh
EH1 2HT

Dr P M Taylor
Lecturer
Department of Electronic Engineering
University of Hull
Cottingham Road
Hull
HU6 7RX

Chapter 1

Second generation robotics and robot vision

Professor Alan Pugh

1. Introduction

During the early years of their existence, Industrial Robots represented a solution in search of a problem. At the first International Symposium on Industrial Robots held in Chicago in 1970[1] the delegates were treated to a brief catalogue of robot applications in industries which were invariably hot, smelly and involved jobs requiring a great deal of muscle power. This was the era when the industrial robot was a mere curiosity and its existence was known only to a relatively few informed industrialists. It was the combination of the industrial robot with the problem of spot welding automobile bodies which allowed the versatility of the industrial robot to be properly exploited. This single application transformed the industrial robot scene overnight which resulted in an escalation of the number of robots employed in industry coupled with a liberal coverage in the media which in turn stimulated interest to extend the application of industrial robots to other industrial tasks.

Despite the increasing use of robots in industries having a hostile environment — for example dyecasting and forging — the greater success in application has hitherto been in areas where precise contact with the work place has not been an essential ingredient of the application.[2] In addition to spot welding which falls into this category, paint spraying has been a popular and successful application and considerable success in recent years has been established with seam welding. While the industrial robot was originally conceived with a point-to-point mode of operation, it has been the continuous path mode of operation which has dominated most industrial robot applications. The situation existing now is that robots have become a natural and essential feature for the establishment of a manufacturing process particularly if this is being assembled on a green-field factory site. No longer do potential users need to wait for some adventurous industrialist to experiment with applications for the first time; most of this has been done and applications in the *non-sensory* handling and processing areas are well catalogued. Further, developments in robot structures and mechanisms now give the user a superb range of devices/products from which to choose as any visit to an exhibition of robots will demonstrate.

2. Where is 'generation 2'?

From the very beginning of industrial robot technology, the prospect of intelligent control coupled to environmental sensors has been 'just around the corner'. That 'corner' is now over ten years old and still no really satisfactory realisation of the 'second generation' has been forthcoming. The impressive and promising demonstrations of 'artificial intelligence' at the First International Symposium[1] gave way to more relevant realisations of sensory interaction during the 1970's and a few commercial vision systems are now available.[3,4,5] These however, represent the product of research of about a decade ago and demand a high contrast image for reliable operation.

Cost too plays its part as a deterrent in the sluggish approach of the 'second generation'. We see again the natural reluctance of industrial users not wishing to be first in the field. Acknowledging that substantial developments might exist behind the 'closed doors' of industrial confidentiality, only a few brave experiments towards the 'second generation' have appeared in the technical press. Incentive plays its part in the use of sensory control for certain applications and the microelectronic industry has provided a need for such radical thinking. Examples of this work have been published for some time.[6,7] The need for vision in this industry is associated with the high degree of visual feedback required during the act of device fabrication and assembly and the substantial geometric content of integrated circuit pellets has encouraged the application of machine vision to the problem of automatic alignment and wire bonding. The important aspect of this application which has demonstrated reliable operation must be the relative ease of control of the workplace and the illumination of the assembly area.

Can this success in the use of sensory control be regarded as a realisation of the 'generation 2' robot? Some would argue that it does but most would not accept the specialist mechanised handling as representative of versatile 'programmable automation'. Published attempts of sensory control of robots for shop-floor application are difficult to find. Single experimental applications of both visual and tactile feedback have been implemented although we have an example in the United Kingdom of a company marketing as a standard feature a vision controller for paint spraying[8] which is supported by several years of operating experience.

The application areas for the first generation robots are rarely associated with automated assembly excepting the applications of non-programmable placement devices (or robots). The breakthrough required to extend the application of programmable robots into automated assembly which will open the door to flexible manufacturing systems can only be realised with sensory control to support the assembly process. Despite recent developments in robot architecture,[9] no programmable robot offers the kind of positioning precision to permit reliable assembly to take place without innovations of sensory control of the simplicity of the 'remote centre compliance'.[10] Automated programmable assembly coupled with sensory interactive handling represents the goal which best defines 'generation 2' robotics. The manufacturing areas which involve small to medium sized batches dominate in the industrial scene and those are the areas which are starved of auto-mated solutions. To achieve success in sector, a great deal of work is still required before the flexibility offered by highly expensive 'generation 2' devices can be justified.

3. Vision research

It is perhaps vision more than the senses of touch and hearing which has attracted the greatest research effort. However, robot vision is frequently confused with vision applied to automated inspection and even the artificial intelligence aspects of scene analysis. If an uninformed comparison is made between the technology of picture processing and the requirements of robot vision it is not possible to reconcile the apparent divide which exists between the two.

The essential requirements for success in robot vision might be summarised as follows:

> low cost
> reliable operation
> fundamental simplicity
> fast image processing
> ease of scene illumination

These requirements are often diametrically opposed to the results of research effort published by research organisations. The processing of grey-scale images at high resolution often provides impressive results but inevitably this is achieved at the expense of processor architecture and processing time. Dedicated image processing systems[11] will attack the problem of processing speed in a most impressive way but there is often a desire on the part of many researchers to identify an area of technological challenge in image processing to satisfy their own research motivation rather than attempt a simplification of the imaging problems.

Probably the single aspect which causes difficulty but often overlooked is the control of illumination of the work area. This problem has been attacked by some researchers using 'planes of light'[12,13] which might be regarded as a primitive application of structured lighting i.e. super-imposing on the work area a geometric pattern of light which is distorted by the work pieces. The success of this approach is manifested in the simplicity of binary image processing and a reduction in the magnitude of visual data to be analysed.[14] Developments of early demonstrations of robot vision using back lighting of the work area have reached a stage of restricted industrial application.[15] However, a feature of sensory techniques which have industrial potential, is that they are often application specific and cannot be applied generally.

The experiments linking image processing of televised images with robot applications[16] (for example) are in their infancy at present; the cost is high and the reliability of image processing is unlikely to be satisfactory for some time to come.

4. Image resolution

Sensing for robot applications is not dependent on a relentless pursuit for devices with higher and higher resolution. The fundamental consideration must be in the selection of *optimum* resolution for the task to be executed. There is a tendency to assume that the higher the resolution then the greater is the application range for the system. At this point in our evolution, we are not exploiting the 'state of the art' as much as we should. Considerable success has been achieved using a resolution as low as 50 × 50.[6] With serial processing architectures, this resolution will generate quite sufficient grey-scale data to test the ingenuity of image processing algorithms! Should processing time approach about 0.5 seconds, this will be noticeable in a robot associated with handling. However, for welding applications, the image processing time must be even faster.

High resolution systems are required in applications involving automated inspection and picture data retrieval where speed is sometimes not such an important criterion. This must not be confused with the needs of sensory robot systems.

5. The sensor crisis

Perhaps the key issue in the production of the sensory robot is the availability of suitable sensors. The following represents a summary of sensing requirements for robot applications:

> presence
> range
> single axis measurement (or displacement)
> 2-dimensional location/position
> 3-dimensional location/position
> thermal
> force

for which the following sensing devices or methods are available.

Vision	*Acoustic*
Photo-detector	Ultrasonic detectors/emitters
Linear array	Ultrasonic arrays
Area array	Microphones (voice control)
T.V. camera	
Laser (triangulation)	*Tactile*
Laser (time of flight)	Probe
Optical fibre	Strain gauge
	Piezoelectric
Other	Carbon materials
Infra red	Discrete arrays
Radar	Integrated arrays
Magnetic proximity	
Ionising radiation	

The only satisfactory location for sensors is on the robot manipulator itself at or near the end effector.[14,17] Locating an image sensor above the work area of a robot suffers from the disadvantage that the robot manipulator will obscure its own work area and the metric displacement of the end effector from its destination must be measured in an absolute rather than a relative way. Siting the sensor on the end effector allows relative measurements to be taken reducing considerably the need for calibration of mechanical position and the need for imaging linearity. Sensory feedback in this situation can be reduced to the simplicity of range finding in some applications.

What is missing from the sensor market are devices specifically tailored to be integrated close to the gripper jaws. The promise of solid-state arrays for this particular application has not materialised which is primarily due to the commercial incentives associated with the television industry. It might be accurate to predict that over the next decade imaging devices manufactured primarily for the television market will be both small and cheap enough to be useful in robot applications. However, at present, area array cameras are extremely expensive and, while smaller than most thermionic tube cameras, are far too large to be installed in the region of a gripper. Most of the early prototype arrays of modest resolution have been abandoned.

It is not an exaggeration to suggest that no imaging sensors exist which are ideally suited for robot applications. The use of dynamic RAM devices for image purposes[17] has proved to be a minor breakthrough and gives an indication of the rugged approach which is needed to achieve success. Some researchers have attacked the problem of size reduction by using coherent fibre optic to retrieve an image from the gripper area[16] which imposes a cost penalty on the total system. This approach can, however, exploit a fundamental property of optical fibre in that a bundle of coherent fibres can be subdivided to allow a single high-resolution imaging device to be used to retrieve and combine a number of lower resolution images from various parts of the work area including the gripper — with each subdivided bundle associated with its own optical arrangements.[18,19]

Linear arrays have been used in situations involving parts moving on a conveyor in such a way that mechanical motion is used to generate one axis of a 2-dimensional image.[12,18] There is no reason why the same technique should not be associated with a robot manipulated by using the motion of the end effector to generate a 2-dimensional image. Also implied here is the possibility of using circular scanning of the work area or even taking a stationary image from a linear array.

Tactile sensing is required in situations involving placement. Both active and passive compliant sensors[20,10] have not only been successfully demonstrated but have experienced a period of development and application in the field. The situation surrounding tactile array sensors is quite different. Because the tactile arrays are essentially discrete in design, they are inevitably clumsy and are associated with very low resolution. Interesting experiments have been reported[21,22,23,24] and an exciting development for a VLSI tactile sensors is to be published.[25]

Experiments with range sensing have been liberally researched and some success has been achieved with acoustic sensors and optical sensors (including lasers). The whisker probe[24] can now be replaced by a laser alternative with obvious advantages. Laser range finding is well developed but under-used in robot applications although the use of laser probes sited on the end effector of an industrial robot makes a natural automated dimensional inspection system.

It is clear from this brief catalogue of sensing methods that a great deal of chaos surrounds the sensor world. What we must work towards is some element of modularity in sensor design to allow for the optimum sensing method to be incorporated into a given application. No single sensor can provide the solution to all problems and bigger does not always mean better. However, the recent exciting developments in 'smart sensors' which incorporate primitive image processing (front-end processing) will be most welcome. A comprehensive survey and assessment of robotic sensors has been published by Nitzan.[26]

6. Languages and software

The involvement of the stored-program computer in the present generation of robots does no more than to provide an alternative to a hard-wired controller excepting that the computer provides an integrated memory facility to retain individual 'programs'.

Software and languages became a reality for most users with the introduction of VAL[27] which incorporates the capability to interpolate linear motion between two points and provides for co-ordinate transformation of axes. Further, VAL allows for transformation between vision and machine co-ordinates.[28] Machine training or learning using VAL can be achieved 'off-line' rather than the 'teach by showing' which is the method used by the majority of present generation robots.

Work is now under way on languages for assembly[29,30] which will give to the sensory robot system autonomy of action within the requirements of the assembly task. Further, a common assembly language will permit the same instructions to be repeated on different robot hardware in the same way that computer programs written in a common language can be executed on different machines. The lesson to be learned here is that we must discipline ourselves to a common assembly language before a proliferation of languages creates a situation disorder. It is still early days to consider an assembly task being executed by an 'off line' assembly language as part of a CAD/CAM operation. One of the stepping stones required in this revolutionary process is the establishment of some 'bench marks' to compare and test the relative merits of assembly languages.

7. Concluding remarks

The ingredients which comprise the second generation robot are:

mechanisms with speed and precision
cheap and reliable sensors
elegant and rugged software

In all of these areas there exists a significant deficiency of development and perhaps a need for an innovative breakthrough. We have seen previously the shortfall in sensor requirements and the need for good supporting software has only been admitted in recent times. With the exception of mechanisms specifically designed for dedicated tasks, no existing robot device alone can really provide the precision necessary for assembly operations. A promising way to proceed is to use a programmable high speed manipulator for coarse positioning, coupled with a 'floating' table which incorporates features for fine positioning.[31]

When it is remembered that *research* demonstrations of 'generation 2' robots have been available over the past decade[32,33,34,35,36,37] it is salutory to recognise that predictions for the future made over this period have not been realised. A survey of the current situation in robot vision has been published recently.[38] It would be a brave person who now predicted where and when the 'generation 2' robot would take its place in industry. With the wisdom of hindsight we know that there is still a vast amount of research and development required coupled with an industrial need for a 'generation 2' robot. Perhaps this need will first appear in the textile, pottery or confectionery industries[39] to provide the 'shot in the arm' for 'generation 2' just as spot welding did for 'generation 1' a decade ago.

Over the past two years, the United Kingdom has introduced government funding to aid and support industrial applications as well as providing a co-ordination programme of research and development in the universities. Surveys of industrial requirements and research partnerships in the U.K. have recently been published.[40,41]

8. References

1. First International Symposium on Industrial Robots. Illinois Institute of Technology, Chicago, April 1970.
2. ENGELBERGER, J. F. 'Robotics in Practice' Kogan Page, London, 1980.
3. GLEASON, G. J. and AGIN, G. J. 'A Modular Vision System for Sensor-controlled Manipulation and Inspection'. 9th International Symposium on Industrial Robots, SME, Washington, March 1979.
4. HEWKIN, P. F., and PHIL, M. A. D., 'OMS – Optical Measurement System' 1st Robot Vision and Sensory Controls, IFS (Conferences) Ltd, Stratford, England, April 1981.
5. VILLERS, P., 'Present Industrial Use of Vision Sensors for Robot Guidance' 12th Industrial Symposium on Industrial Robots, l'Association Francaise de Robotique Industrielle, Paris, June 1982.
6. BAIRD, M. L., 'SIGHT-1: 'A Computer Vision System for Automated IC Chip Manufacture' *IEEE Trans. Systems, Man and Cybernetics,* Vol. 8, No. 2, 1978
7. KAWATO, S. and HIRATA, Y., 'Automatic IC Wire Bonding System with T.V. Cameras' SME Technical Paper AD79–880, 1979.
8. JOHNSTON, E., 'Spray Painting Random Shapes Using CCTV Camera Control', 1st Robot Vision and Sensory Controls, IFS (Conferences) Ltd, Stratford, England, April 1981.
9. MAKINO, H., FURUYA, N., SOMA, K., and CHIN, E., 'Research and Development of the SCARA robot', 4th International Conference on Production Engineering, Tokyo, 1980.

10. WHITNEY, D. E., and NEVINS, J. L., 'What is the Remote Centre Compliance (RCC) and What Can it do?' 9th International Symposium on Industrial Robots, SME, Washington, March 1979.

11. FOUNTAIN, T. J., GEOTCHERIAN, V., 'CLIP 4 Parallel Processing System', *IEE Proc.,* Vol. 127, Pt. E, No. 5, 1980.

12. HOLLAND, S. W., ROSSOL, L., and WARD, M. R., 'Consight-1: A Vision Controlled Robot System for Transferring Parts from Belt Conveyors'. Computer Vision and Sensor Based Robots, Dodd, G. G., and Rossol, L., Eds. Plenum Press, New York, 1979.

13. BOLLES, R. C., 'Three Dimensional Locating of Industrial Parts', 8th NSF Grantees Conference on Production Research and Technology, Stanford Calif., Jan. 1981.

14. NAGEL, R. N., VENDER-BRUG, G. J., ALBUS, J. S., and LOWENFELD, E., 'Experiments in Part acquisition Using Robot Vision', SME Technical Paper MS79–784 1979.

15. SARAGA, P., and JONES, B. M., 'Parallel Projection Optics in Simple Assembly', 1st Robot Vision and Sensory Controls', IFS (Conferences) Ltd, Stratford, England 1981.

16. MAZAKI, I., GORMAN, R. R., SHULMAN, B. H., DUNNE, M. J., and TODA, H., 'Arc Welding Robot with Vision', 11th Industrial Symposium on Industrial Robots, *JIRA,* Tokyo, October 1981.

17. TAYLOR, P. M., TAYLOR, G. E., KEMP, D. R., STEIN, J., and PUGH, A., 'Sensory Gripping System': The Software and Hardware Aspects' *Sensor Review,* October, 1981.

18. CRONSHAW, A. J., HEGINBOTHAM, W. B., and PUGH, A., 'Software Techniques for an Optically-tooled Bowl Feeder', IEE Conference 'Trends in On-line Computer Control Systems', Sheffield, England, Vol. 172, March 1979.

19. STREETER, J. H., 'Viewpoint – Vision for Programmed Automatic Assembly', *Sensor Review,* July 1981.

20. GOTO, T., et al., 'Precise Insert Operation by Tactile Controlled Robot', *The Industrial Robot,* Vol. 1, No. 5, Sept. 1974.

21. LARCOMBE, M. H. E., 'Carbon Fibre Tactile Sensors', 1st Robot Vision and Sensory Controls, IFS (Conferences) Ltd., Stratford, England, April 1981.

22. PURBRICK, J. A., 'A Force Transducer Employing Conductive Silicon Rubber' Ibid.

23. SATO, N., HEGINBOTHAM, W. B., and PUGH, A., 'A method for three-dimensional Robot Identification by Tactile Transducer', 7th International Symposium on Industrial Robots, *JIRA,* Tokyo, October, 1977.

24. WANG, S. S. M., and WILL, P. M., 'Sensors for Computer Controlled Mechanical Assembly', *The Industrial Robot,* March 1978.

25. TANNER, J. E., and RAIBERT, M. H., 'A VLSI Tactile Array Sensor', The 12th International Symposium on Industrial Robots, l'Association Francaise De Robotique Industrielle, Paris, June 1982.

26. NITZAN, D., 'Assessment of Robotic Sensors', 1st Robot Vision and Sensory Controls, IFS (Conferences) Ltd., Stratford, England, April 1981.

27. 'Users Guide to VAL – Robot Programming and Control System', Unimation Inc., Danbury, Conn.

28. CARLISLE, B., 'The PUMA/VS-100 Robot Vision System', 1st Robot Vision and Sensory Controls, IFS (Conferences) Ltd., Stratford, England, April 1981.

29. LEIBERMAN, L. I., and NELSEY, M. A., 'AUTOPASS – An Automatic Programming System for Computer Controlled Mechanical Assembly', *IBM Journal of Research and Development,* Vol. 21, No. 4, 1977.

30. POPPLESTONE, R. J., et al 'RAPT: A language for Describing Assemblies', University of Edinburgh, U.K., Sept. 1978.

31. HOLLINGUM, J., 'Robotics Institute Teams Development in University and Industry', *The Industrial Robot,* December 1981.

32. HEGINBOTHAM, W. B., GATEHOUSE, D. W., PUGH, A., KITCHIN, P. W., and PAGE, C. J., 'The Nottingham SIRCH Assembly Robot', 1st Conference on Industrial Robot Technology IFS Ltd., Nottingham, England, March 1973.

33. TOSUBOI, Y., and INOUE, T., 'Robot Assembly System Using TV Camera' 6th International Symposium on Industrial Robots' IFS Ltd., Nottingham, England, March 1976.

34. ROSEN, C., et al 'Exploratory Research in Advanced Automation', Stanford Research Institute NSF Report, August 1974.

35. SARAGA, P., and SKOYLES, D. R., 'An Experimental Visually Controlled Pick and Place Machine for Industry', 3rd International Conference on Pattern Recognition, IEEE Computer Society, Coronado, Calif., Nov. 1976.

36. ZAMBUTO, D. A., and CHANEY, J. E., 'An Industrial Robot with Mini-Computer Control', 6th International Symposium on Industrial Robots, IFS (Conferences) Ltd., Nottingham, England, March 1976.

37. BIRK, J. R., KELLEY, R. B., and MARTINS, H. A. S., 'An Orientating Robot for Feeding Workpieces Stored in Bins', *IEEE Trans. Systems Man and Cybernetics,* Vol. 11, No. 2, 1981.

38. KRUGER, R. P., and THOMPSON, W. B., 'A Technical and Economic Assessment of Computer Vision for Industrial Inspection and Robotic Assembly' *Proc. IEEE,* Vol 69, No. 12, 1981.

39. CRONSHAW, A. J., 'Automatic Chocolate Decoration by Robot Vision', 12th International Symposium on Industrial Robots, l'Association Francaise de Robotique Industrielle, Paris, June 1982.

40. KING and LAU 'Robotics in the U.K.', *The Industrial Robot,* March 1981.

41. DAVEY, P. G., 'U.K. Research and Development in Industrial Robots' 2nd International Conference on 'Manufacturing Matters' The Institution of Production Engineers, London, England, March 1982.

Chapter 2

Research and development in the United Kingdom

P. G. Davey

Most publicly funded UK research is supported by the Science and Engineering Research Council (Department of Education and Science) and the Department of Industry: the main institutions concerned, and their areas of work are summarised in Appendicies 1 and 2 respectively. There is, however, significant and growing Defence robotics activity not reported here for security reasons; this mainly centres on incorporating more autonomy in remote-guided vehicles.

In June 1980 the SERC launched a major new research programme, in Universities and Polytechnics, with the objective of developing the technology required for 'second generation' robots able to cope accurately with small deviations in their surroundings. The programme is also aimed at bridging two serious gaps in UK engineering culture today:

(a) Between engineers in academia and in industry

and

(b) Between 'information engineering' and production engineering

The Policy in this programme is to make larger awards conditional on the formation of partnerships between academic groups and firms applying, or building, robotic equipment. This must be done before the project even starts so that targets are jointly defined. Trying to do so some years into the project often seems a recipe for failure. We believe that as well as improving graduate training, good academic research in robotics can indeed provide valuable longer-term results for industry, provided that firms participate actively right from the beginning. We require that the joint project be an important part of their corporate strategy agreed at board level, and the firm's total contribution (including a participating engineer) in fact varies from 20 to 68 percent in the schemes listed in Appendix 1. Direct SERC support of this work now totals about £1M p.a. (overhead support from normal University funds adds at least 70%), including salaries of some 70 research workers: comparable support of academic groups in Japan appears to be not less than £3.7M (270 staff). However, the contrast is much more marked when we remember that most robotics R and D is funded not publicly, but privately within Japan's industrial laboratories, and relatively far more than here!

The SERC programme is centrally co-ordinated: the research projects whose 'centre of gravity' is classified in Appendix 1 have been deliberately directed to ensure that most areas (and applications) of need are covered. In areas of particular importance several groups are supported, but the nature and extent of any overlap is carefully considered. From 1982, the programme is entering a second phase in which the emphasis will be on close monitoring of all groups, and concentration of support on those showing the best 'track record'. At present, the 22 major partnership projects are taking place within 17 different institutions: it would seem desirable eventually to concentrate the major projects within ten or so stronger institutions.

Work on language for off-line programming is considered particularly important to many firms interested in robotics and the need to avoid duplication is paramount. Under 4.1 in the Appendix it will be noticed that this research is being steered by a national Language Working Party. This group comprises 5 industry, 4 academic and one NEL member; it has made research contracts with University groups at Edinburgh and Cranfield (their work is also linked to several centres in France and Germany). It is hoped that a standard may eventually emerge from this work, and direct support for associated 'downstream' development work is now being sought from interested firms.

A summary of the research goals for each SERC project listed in Appendix 1 is available[1] and is periodically up-dated. Space does not permit a fuller description of these here, nor mention of many smaller robotics research projects in Universities which are supported from 'background' UGC funding.

Training of future engineers at the post graduate level is an equally important part of SERC's remit: it lies outside the scope of this paper but training in robotics is seen as combining the greatest importance to industry with an attractive mixture of nearly all engineering disciplines. This is reflected in the robotics content in several of the SERC/DoI joint Teaching Company projects, concerned simultaneously with training post-graduate Associates and introducing modern manufacturing techniques into the host firm. Cranfield, Hull and Surrey have set an early lead with MSc and 'short' courses; UMIST, Birmingham, UWIST and Imperial College are expected to follow soon. A number of Universities have also incorporated substantial robotics options and projects in undergraduate courses.

Appendix 2 is a summary of R & D in industrial robotics at DoI Laboratories and the research associations. At PERA, perhaps the most important work being carried out is the Robot Advisory Service. This is a major publicly-funded attempt to provide the crucial applications engineering, without which R & D aimed at second generation robots will be fruitless. So far 170 visits to Companies have been undertaken by PERA to discuss specific projects, and a large number of specialist discussion group meetings as well as seminars on industrial robot technology have been held. 25 detailed feasibility studies have been completed. While detailed information on these studies cannot be given because of confidentiality restrictions, it is noteworthy that in a set of 13 possible applications, two involving one robot each had calculated pay-back periods of 13 and 14 months respectively. Both were parts handling applications, one in heat treatment and the other in conjunction with a bending brake, and both required a general purpose gripper to cope with variations in the workpiece, showing the importance of attempts to design the gripper and fixtures in a way that matches the flexibility of the robot itself.

Experience has shown that short payback periods are indeed possible in some cases like these, but in general it should be borne in mind that robots are essentially flexible tools which should therefore be regarded as economic even if the payback period is much longer.

There is also active development work at PERA which includes projects on industrial vision linked to flexible feeders, and the well known project on graphics simulation being done as a partnership with the SERC supported group at Nottingham University. A semi-robotic welding system has also been developed for the Joint European Torus project comprising a small robotic welder captive inside the structure which can weld structural joints within a large diameter torus. It can also be used for maintenance and repairs during the lifetime of the system.

The National Engineering Laboratory (NEL) is the Department of Industry laboratory with the most extensive robotics research and development. The projects here are part of the DoI's national programme on automated small batch production (ASP). They include an automated FMS for matching prismatic components; an automated assembly and test rig for small electric motors using a Unimate PUMA; and an industrial vision system for parts inspection and control of robotic devices. NEL are also developing an automatic turning cell and a force sensing chuck as part of this programme. As at PERA, the projects are conducted in collaboration with industry and there is additional consultancy work undertaken in robotics and automation under direct contracts with particular firms.

The National Physical Laboratory (NPL) is another major DoI laboratory. It is not working in robotics as such, but has active R & D programs in allied technology, for example, automated 3-dimensional metrology based on laser-generated interference fringes, and close range photogrammetry. The work is expected to have application to robot positioning systems as well as to quality inspection and testing.

The British Cast Iron Research Association (BCIRA) and Steel Castings Research and Trade Association (SCRATA) are both concerned with robot fettling systems. Some castings are of a size which can be held by a robot against different fettling tools — note that this class can now be extended to castings which are normally much too heavy for a man to pick up and manipulate against the tool. Robots are also used for fettling of even heavier castings; here the robot acts as a crude machine tool, moving the tool as required to take off unwanted metal. It is expected that measurement of force and torque and incorporation of this data into control of the robot will be an important part of robotic fettling systems in the future.

At least three other research associations are concerned with particular aspects of robotics although they do not carry out R & D in robotics generally. These are also included in Appendix 2.

Reference

1. DAVEY, P. G., '*SERC Industrial Robotics Initiative: Progress Report*', March 1982.

Appendix 1

1. Manipulators and Dynamic Control
   1.1 Theory: Edinburgh, QMC, Liverpool Poly, Newcastle, OU
   1.2 Software: Surrey/Hall Automation
   1.3 Hardware: Loughborough/Martonair, UCL

2. Generic Software
   2.1 Workplace Planning: Nottingham/PERA
   2.2 User's View: Cambridge
   2.3 Image Handling: QMC/Micro Consultants, Edinburgh/GEC Hirst Royal Holloway College

3. Arc Welding
Loughborough/BR, Newcastle/Auston Pickersgill, Oxford/BL.

4. FM Cells and Assembly
   4.1 Language, and CAD/CAM links: Cranfield, Edinburgh, Language Working Party
   4.2 Vision guidance: Cranfield/Remek, Hull/GEC Marconi.
   4.3 Tactile Guidance: Cranfield/Remek,UWIST (to be announced)
   4.4 Grippers: Birmingham/Unimation, Cranfield/Parkinson Cowan
   4.5 Product Redesign: Cranfield, Salford/Fairey
   4.6 Error Recovery: Aberystwith/BRS

5. Grinding and Polishing: Bath/Walker Crossweller

6. Unmanned Industrial Trucks: Warwick/Lansing Bagnall

7. Textiles: Hull/Corah, Loughborough/Coates Patons, Durham/Lyall and Scott

8. Composites: QUB/Short Bros

9. Robotic Inspection and Calibration: Lanchester Poly, Loughborough/LK, Portsmouth Polytechnic/TI.

APPENDIX 1:   Functional Breakdown of R & D in Industrial Robotics supported by SERC
         (Major Partnerships shown thus: University/Firm)

Appendix 2

Production Engineering Research Association (PERA) Robot Advisory Service:

Industrial Vision, Flexible Feeders, Graphics Simulation, Semi-Robotic Welding

National Engineering Laboratory (NEL):
FM Systems, Automated Assembly, Industrial Vision, Automatic Turning Cell.

National Physical Laboratory (NPL):
3-D Metrology, close-range photogrammetry, advanced automated inspection.

British Cast Iron Research Association (BCIRA): Steel Castings Research and Trade Association (SCRATA):
Robot fettling systems and handling of castings.

Machine Tool Industry Research Association (MTIRA) (with Health and Safety Executive):
Standardisation, safety, and guarding.

Scientific Instrument Research Association (SIRA):
Sensory systems for safety and navigation, low cost "smart" sensors.

Drop Forging Research Association
Manipulators and robots for forging.

APPENDIX 2: Non-Proprietary R and D in Industrial Robotics in DoI Laboratories and Research Associations

Chapter 3

The robot control problem

J. R. Hewit

1. The control problem

The design of a control system for a robotic machine can be very straightforward or can be extremely difficult depending on a number of factors. These are

(a) the geometric design of the machine
(b) the specifications of speed, accuracy etc
(c) the degree of definition of the task — i.e. how well structured the task is
(d) the presence, or not, of external disturbances
(e) the degree to which the machine may be modelled dynamically
(f) the method of teaching the machine

These factors are of course inter-related so that, for instance (a) will have a major influence on (b), and so on.

In the simplest application of 'robots', which might be taken to be 'pick and place' operations in which the machine merely groups an object, moves it to another position, releases it and then repeats, control is effected by the prior placement of fixed end-stops and actuator is provided by (usually) pneumatic cylinders which drive the degrees of freedom up against the stops in a prerecorded sequence.

This sequence is 'recorded' in some simple way which may be mechanical, pneumatic, fluidic or electronic. 'Robots' of this type can clearly have use only in highly structured types of task.

At the other end of the control spectrum are the advanced computer controlled arms which are now being designed for tasks such as automatic assembly.

In such machines, full coordinated control of each degree of freedom is to be effected by sophisticated control algorithms which will take into account the kinematics and dynamics of the system and compute the forces and torques necessary to drive the end-effector along desired trajectories in the working space.

The control computers will have access to sensory information, of forces encountered in accomplishing the task, for example, or visual images which may be processed to provide knowledge of the environment. The control algorithms may be required to make use of this information in generating the demand signals for the limb servo's.

From this it may be appreciated that one of the severest constraints imposed upon the control system is the very short time available in which to perform the algorithm if this must be accomplished in real-time.

In control theoretic terms, the robotic system in this severest case, may be described as multivariable, interacting, non-linear, time-varying, and partially modelled.

No applicable corpus of control theory exists to deal with systems possessing such a combination of problematic features and today research is still at a relatively early stage in seeking solutions. However, a number of promising avenues are being explored and a few proposals have been made from which might be expected to emerge a new generation of 'intelligently controlled' robots.

2. Kinematic considerations

At the heart of any robotics application lies the kinematic configuration of the machine itself. This will be dictated by the machine designer in meeting the specifications laid down by whosoever commissioned the design. This configuration, once established, will then define limits to the set of tasks to which the machine may be applied.

Although very many kinematic configurations are possible the vast majority of robot arms have designs which fall into one of three categories.

These are (a) Rectangular
          (b) Cylindrical
          (c) Spherical

and are named for the coordinate system which is natural to the movement of the end-effector, or tool, in 3-D space. (Fig. 1).

Broadly speaking applications of these three categories are as follows:-

(a) precision assembly
(b) machine tool loading
(c) general transfer, welding.

In addition to these three degrees of freedom (d-o-f) the end-effector will itself possess a number of degrees of freedom to permit its orientation in space. In most cases this will add three more d-o-f since 6 d-o-f are required to define the position and orientation of a body in space.

Again, broadly speaking, these 6 d-o-f will be provided by 6 single d-o-f mechanical joints of only two types — prismatic or rotary, and each will be driven by a motor via some suitable gear and transmission.

If we denote by $\theta_1, \theta_2, \ldots \theta_6$ the six joint displacements (angular or linear) and by $x_1, x_2 \ldots x_6$ the six coordinates of position and orientation of the end-effector in space then clearly there exists a relationship of the form (in matrix notation).

$$\mathbf{X} = \mathbf{X}(\boldsymbol{\theta}) \tag{1}$$

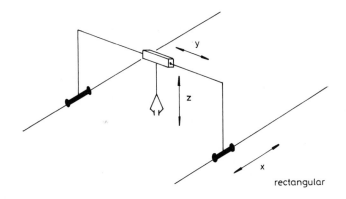

rectangular

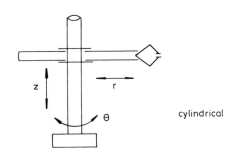

cylindrical

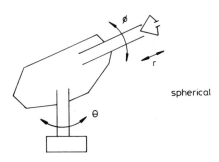

spherical

Fig. 3.1

which indicates that if the robot's joints are given definite displacements then the end-effector will thereby be given a definite position and orientation in space, this of course being the function of the machine.

Usually **X** is called the *world-coordinate vector* and **θ** the *joint-coordinate vector* (a third system of coordinates may also be useful in some tasks – the *tool-coordinate vector* which defines displacements in a reference frame attached to the end-effector).

It should be noted here that the joint displacement θ_i is not in general the same as the displacement of the motor which actuates joint *i*. Depending upon the transmission system used, the relationship between θ_i and the displacement of the *i*th joint motor may or may not be linear. (See Fig. 2). Thus if a tachogenerator is attached to the motor shaft to provide velocity feedback some transformation may be necessary to extract the joint velocity $\dot{\theta}_i$.

In most cases, eqn. (1) represents a set of highly non-linear equations and, depending upon the method of teaching the robot and on the degree of structure inherent in the task, may or may not cause problems in control system design.

In many applications, such as spraying or arc-welding, the robot is *'taught-by-doing'*. Here the operator puts the robot into 'teach' mode and then moves the arm (or an unpowered master) around the required trajectory. During this period the machine records the time-variation of its joints and stores this data in memory. When it is desired to perform the actual task the operator puts the robot into 'repeat' mode, the robot retrieves the trajectory data from memory and exactly repeats the taught movements. In this case, the kinematic equations need not be solved and pose no problem.

In many applications, however, teaching by doing is not appropriate. Where the robot is to be incorporated into a versatile automation system it might be that a description of the task and its associated trajectories would be passed to a control computer via some operators program. Thus for example part of a hypothetical robot control program might contain the statements

> GRASP
> MOVE TO 100, 200, 150
> RELEASE

which are intended to cause the end-effector to grasp an object, move it to a point distant 100 mm in the *x*-direction, 200 mm in the *y*-direction and 150 mm in the *z*-direction and then release it.

In such cases (and increasing numbers of robots becoming commercially available possess some such language) the control system must deal with the kinematic equations.

A number of different strategies for kinematic control are available.

The direct method is to invert eqn. (1) to obtain the arm-coordinates thus

$$\mathbf{\theta} = \mathbf{\theta}\,(\mathbf{X}) \tag{2}$$

For robot arms with very simple kinematic configurations this inversion may be

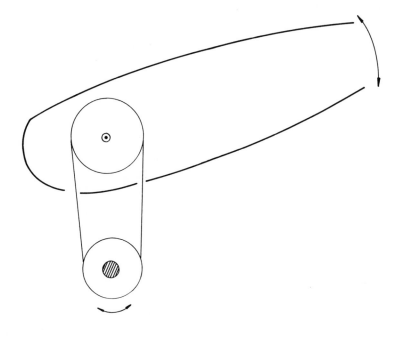

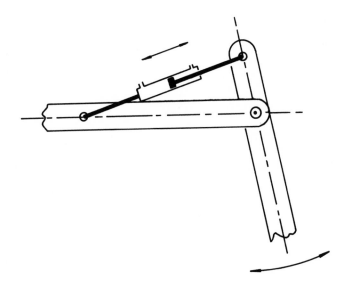

Fig. 3.2

accomplished analytically but in most cases some computational algorithm is required, probably of a recursive nature and hence time-consuming. For arms which possess a degree of redundancy, i.e. where the number of arm-coordinates is greater than 6, an infinite number of solutions exists and some stratagem must be involved to select one of these.

Assuming, however, that the inversion can be accomplished in real time or that the trajectory can be planned ahead of time, the control system will have the form shown in Fig. 3.

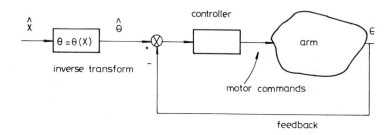

Fig. 3.3.

That this world-to-joint coordinate transformation is by no means trivial may be better understood by consideration of the computations involved for a typical robot. For a six-joint Unimate — which has kinematic features which *ease* the computational burden — the transformation requires, at least, the following

Multiply/Divide	30
Add/Subtract	15
Square Root	3
Square	7
Cosine/Sine	6
Arctangent	6

Since it is necessary to output commands to the arm servos at a frequency of at least 50/sec it is clearly infeasible to implement such a control scheme in real-time and other methods must be employed. The most significant of these is Resolved Motion Rate Control.

3. Resolved motion rate control

If eqn. (1) is differentiated once we have

$$\dot{X} = J(\theta) \cdot \dot{\theta} \tag{3}$$

Here $\mathbf{J}(\boldsymbol{\theta})$ is called the *Jacobian matrix* of the system whose typical element J_{ij} is given by

$$J_{ij} = \frac{\partial x_i}{\partial \theta_j} \tag{4}$$

and defines by how much the ith world coordinate changes for a given change in jth arm coordinate.

For a non-redundant robot arm of non-pathological design the matrix $\mathbf{J}(\boldsymbol{\theta})$ is non-singular and may be inverted so that it is possible to solve the inverse velocity equation

$$\dot{\boldsymbol{\theta}} = \mathbf{J}^{-1}(\boldsymbol{\theta}) \cdot \dot{\mathbf{X}} \tag{5}$$

If we wish the overall system to be decoupled so that a single input signal controls one and only one world coordinate it would clearly be necessary to make the world coordinate velocity vector $\dot{\mathbf{X}}$ proportional to the world coordinate error vector $(\hat{\mathbf{X}} - \mathbf{X})$ or

$$\dot{\mathbf{X}} = \mathbf{K}(\hat{\mathbf{X}} - \mathbf{X}) \quad (\mathbf{K} \equiv \text{diag}) \tag{6}$$

or combining eqns. (5) and (6)

$$\dot{\boldsymbol{\theta}} = \mathbf{J}^{-1}(\boldsymbol{\theta})\mathbf{K}(\hat{\mathbf{X}} - \mathbf{X}) \tag{7}$$

The overall control system is shown in Fig. 4. The computational burden involved in this control technique is still, however, by no means negligible.

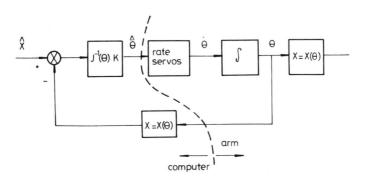

Fig. 3.4.

The angle transducers must be sampled to provide the angle vector $\boldsymbol{\theta}$ and the transformation $X = X(\theta)$ must be computed, the matrix $\mathbf{J}^{-1}(\boldsymbol{\theta})$ must be evaluated and multiplied by the desired gain matrix $\mathbf{K}$.

In practice since the computation of $\mathbf{J}^{-1}(\boldsymbol{\theta})$ would still be too cumbersome to be done in real-time, a constant value for $\mathbf{J}^{-1}(\boldsymbol{\theta})$ could be used over a number of sampling periods and updated only relatively slowly. The degradation in coordination so induced could be made as small as necessary by implementing high gain servo loops by attributing high value to the elements of the **K** matrix. The problem of computing $X = X(\theta)$ could be fairly small particularly if the angle transducers were to incorporate sine and cosine encoders and might readily be solved using a separate microprocessor for each element.

However, it must be reiterated that control schemes based upon this degree of coordinate sophistication are relatively scarce and most commercially available robots rely to some degree of teaching by doing or trajectory planning.

Having considered the implications, for control, of the complex kinematics involved in manipulator structure, it is next necessary to consider dynamical effects.

4. Dynamical considerations

One of the major reasons for the reluctance of industry to employ robotic devices in tasks such as simple assembly is that present day arms are too slow in their movements.

However, it is not possible to increase their speed of operation simply by increasing the power of their motors and demanding higher speeds from the servo's.

As operational speeds approach human-like speeds (above say 1 m/s) it is no longer possible to view a robot arm as a purely kinematic device. At higher speeds the inertia effects due to the velocities and accelerations of the links comprising the arm assume increasing importance and the motors must provide forces and torques to compensate for these effects. Unfortunately the computation of these compensating forces is so immensely difficult that only a small amount of progress has been made to date.

That compensation for dynamic effects *is* possible however is evidenced by human coordination itself and the fast coordinative skills of athletes, sportsmen and musicians indicate that an articulated mechanism with many d-o-f can be forced to produce desired end-effector movement at high speed. Just how this is done, however, remains unclear and research is continuing to resolve the many questions which this raises.

If dynamic effects are to be catered for then a dynamical model for the system must be established.

To do this, direct appeal to the laws of Newton leads to a formulation of the equations of motion in which appear, as dependent variables, the forces of constraint at the arm joints. The number of equations so derived exceeds the number of d-o-f constraint forces and by the imposition of the trigonometric constraints (which hold the arm together).

A simpler approach is to use Lagrangian techniques, based upon energetic principles, which lead to an irreducible set of equations — one second order differential equation for each arm link. This set of equations can be written in the matrix form.

$$\mathbf{A}(\theta) \cdot \ddot{\theta} \; = \; \mathbf{B}(\theta, \dot{\theta}) + D(\theta)\mathbf{T} \tag{8}$$

where for a six d-o-f arm $A(\theta)$ is the 6×6 inertia matrix, $\mathbf{B}(\mathbf{\theta}, \dot{\mathbf{\theta}})$ is a 6×1 vector containing all forces other than reactive and applied, including frictional terms, Coriolis forces, Euler forces and so on. $\mathbf{D}(\mathbf{\theta})$ is a 6×6 matrix whose elements are defined by the placement of the motors in the geometries of the arm. **T** is the vector of motor forces and torques.

(It should be noted, in passing, that since the inertia matrix $A(\mathbf{\theta})$ is non-diagonal the equations cannot be directly solved, for example, by analogue computation).

We must now consider not only the dynamic equations but also the equations which define the kinematic configuration of the arm. To this end differentiate eqn. (1) twice to obtain

$$\dot{\mathbf{X}} = \mathbf{J}(\mathbf{\theta})\dot{\mathbf{\theta}} \tag{9}$$

$$\ddot{\mathbf{X}} = \mathbf{J}(\mathbf{\theta})\ddot{\mathbf{\theta}} + \mathbf{L}(\mathbf{\theta}, \dot{\mathbf{\theta}}) \tag{10}$$

Combining eqn. (10) with eqn. (8) leaves, (ignoring arguments)

$$\mathbf{AJ}^{-1}\mathbf{X} = \mathbf{AJ}^{-1}\mathbf{L} + \mathbf{B} + \mathbf{DT} \tag{11}$$

Now suppose that we try to decouple the overall system by the use of diagonal feedback and by choosing a control matrix $K(s)$ in such a way as to render diagonal the relationship between $\ddot{X}$ and $(\hat{X} - X)$ so that

$$\ddot{\mathbf{X}} = K(s)(\hat{\mathbf{X}} - \mathbf{X}) \equiv K(s)\mathbf{E} \tag{12}$$

where $K(s)$ is some desired diagonal matrix of transfer functions.

Then from eqns. (12) and (11) we evidently require

$$\mathbf{AJ}^{-1}\mathbf{KE} = \mathbf{AJ}^{-1}\mathbf{L} + \mathbf{B} + \mathbf{DT} \tag{13}$$

or $$\mathbf{T} = \mathbf{D}^{-1}\{\mathbf{AJ}^{-1}(\mathbf{KE} - \mathbf{L}) - \mathbf{B}\} \tag{14}$$

Clearly since the satisfaction of eqn. (12) is implied, the overall system reduces to

$$\frac{X}{\hat{X}} = \frac{K(s)/s^2}{1 + K(s)/s^2} = \frac{K(s)}{s^2 + K(s)} \tag{15}$$

In order to render this system stable, one choice for $K(s)$ could be

$$K(s) = k(1 + s\tau) \tag{16}$$

5. Invariance

Let us suppose that a system has two inputs — a signal input and a noise input. The output from the system is required to respond to the signal input only and should ideally contain response to the noise, i.e. the output should be *invariant* with respect to the noise.

One very effective way of achieving this is to measure the noise as it enters the system and to impose upon the signal input an anti-noise component 180° out of phase with the noise. (See Fig. 5).

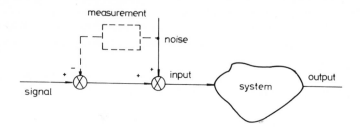

Fig. 3.5.

This concept has been used successfully in the elimination of acoustical noise emerging from engine exhausts and the like.

How this concept may be applied to the control of a complex mechanical system such as a manipulator may be best appreciated by the prior consideration of the control of the linear displacement of a 1 d-o-f mass in the presence of friction (of various forms) and an external disturbance. Figure 6 shows such a system.

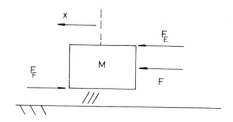

Fig. 3.6.

The equation of motion is clearly

$$M\ddot{X} = F - F_F + F_E \tag{17}$$

where F is the force applied by the motor to control the displacement, F_F is the frictional force and F_E the external force. If we lump F_E and F_F together as

$$P = F_E - F_F \tag{18}$$

which represents a set of $P + D$ controllers in this case, the overall system response becomes

$$\frac{x_i}{\hat{x}_i} = \frac{k(1 + s\tau)}{s^2 + k\tau s + k} \qquad i = 1 - 6 \tag{19}$$

Thus the robot arm behaves as though it consisted of a set of unconnected and non-interacting second order systems whose natural frequencies and damping factors can be set by appropriate choice of controller parameters.

The major and indeed, fatal, drawback to this method of providing simultaneous decoupling and coordinative control lies in the computational complexity of the control algorithm (eqn. 14).

It is extremely unlikely that any combination of fast processing, parallel processing or memory lock-up would be capable of obtaining values of the torque vector in anything approaching real time.

Several ingenious methods to reduce the complexity of this problem have been proposed but it is fair to say that none as yet can be considered even remotely successful and until recently it has appeared that the coordinated control of a robot arm on world-coordinates without trajectory planning might be beyond present-and even foreseeable-future technology.

However, a new technique has emerged based upon a very simple and well known concept in classical control theory. This new technique has such potential that it may not only provide a solution to the robot control problem but it may also give some answers to the questions of how human beings achieve coordinative skills.

The concept upon which this new method is based is *invariance.* We have

$$M\ddot{X} = P + F \tag{20}$$

and following the procedure of the invariance concept we may describe F as the 'signal' input and P as the 'noise' input.

Now in such a mechanical system it is unlikely that the 'noise' P could be directly measured since it is comprised of a number of distinct elements, nor is it feasible to calculate P from measurements of velocity etc since it contains random elements (F_E) as well as elements such as Coulomb friction for which adequate models are unavailable.

It *is* possible and indeed simple however, to obtain an unseable measurement of P indirectly. From eqn. (20) clearly

$$P = M\ddot{X} - F \tag{21}$$

so that if the mass is fitted with an accelerometer to measure $\ddot{X}$, and provided a measurement of the motor applied force F is available and provided the mass M is known then P is calculable from eqn. (21).

If now the applied force is obtained from an actuator with transfer function $W(s)$ which is controlled by two input signals C_C and C_A such that

$$F = W(s)\{C_C - C_A\} \tag{22}$$

then invariance is achieved provided the signal C_A is generated from the calculated P via

$$C_A = \frac{1}{W(s)} \cdot P \tag{23}$$

The overall system then reduces to

$$\frac{X}{C_C} = \frac{W(s)}{Ms^2} \tag{24}$$

which clearly contains no trace of P and to which conventional feedback control techniques can be applied.

It is important to note that since the 'noise' P contains system generated elements i.e. feedback elements such as viscous friction, stability of the system is not guaranteed This problem is further exacerbated by the fact that an *exact* estimate of M is probably unavailable and by the unlikelihood of generating an *exact* inverse of $W(s)$.

Nevertheless, a stability analysis, which is fairly trivial for this simple 1 d-o-f system, shows that provided care is taken in the design of the overall feedback controller adequate margins of stability can readily be obtained.

6. Invariant robot controls

Returning to eqn. (11) the kinematic and dynamic equations of motion for a typical robot arm may be combined and expressed as

$$\mathbf{D}^{-1} \cdot \mathbf{AJ}^{-1} \cdot \ddot{\mathbf{X}} = \mathbf{D}^{-1} \mathbf{AJ}^{-1} \mathbf{L} + \mathbf{D}^{-1} \mathbf{B} + \mathbf{T} \tag{25}$$

or by writing

$$\mathbf{D}^{-1} \mathbf{AJ}^{-1} \equiv \mathbf{I(\theta)} \tag{26}$$

and

$$\mathbf{D}^{-1} \mathbf{AJ}^{-1} \mathbf{L} + \mathbf{D}^{-1} \mathbf{B} \equiv \mathbf{Q(\theta, \dot{\theta})} \tag{27}$$

as

$$\mathbf{I(\theta)} \ddot{\mathbf{X}} = \mathbf{Q(\theta, \dot{\theta})} + \mathbf{T} \tag{28}$$

Comparing eqn. (28) with eqn. (20) it is clear that eqn. (28) is a multivariable, non-linear version of eqn. (20) where $\mathbf{T}$ is the 'signal' and $\mathbf{Q(\theta, \dot{\theta})}$ the noise. The invariance concept described above may thus be applied directly to eqn. (28).

For this to be so it is necessary to equip the robot arm with an array of accelerometers, suitably sited, to measure some acceleration vector $\mathbf{a}$ which may be subsequently processed to provide the world-coordinate acceleration vector $\ddot{\mathbf{X}}$.

The torque vector $\mathbf{T}$ applied by the arm motors must also be available and this can be taken care of, for example, by monitoring the armature current in D.C. motors or by the fitting of force transducers appropriately. An estimate of the matrix $\mathbf{I(\theta)} \equiv \mathbf{D}^{-1} \mathbf{A(\theta)} \mathbf{J}^{-1} \mathbf{(\theta)}$ is also required.

Following the 1 d-o-f invariant control and applying overall conventional feedback round the invariant system we can directly arrive at the control scheme shown in Fig. 7.

Ideally, if all measurements and estimates were exact, then the 'noise' vector $\mathbf{Q(\theta, \dot{\theta})}$ would be rendered ineffective and the robot arm would behave as though it consisted of six unperturbed independent second order systems.

Since it is claimed that such a scheme makes possible the real-time control of a robot arm in world-coordinates it is worthwhile to consider just what computational burden is involved.

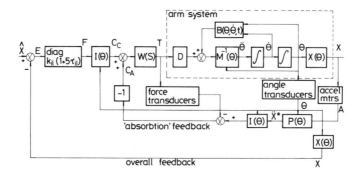

Fig. 3.7.

This falls into 2 main parts.

1. The transformation of the acceleration measurement vector **a** to the world-coordinate accelerometer vector $\ddot{\mathbf{X}}$.

By suitable placement of the accelerometers this can be made to require a relatively simple transformation of the form

$$\ddot{\mathbf{X}} = \mathbf{P(\theta)a} \tag{29}$$

The elements of $\mathbf{P(\theta)}$ contain terms involving sines and cosines of the arm angles and these can be obtained directly from sine and cosine encoders.

Subsequent processing involves a small enough number of multiplications to be handled by a dedicated microprocessor.

2. Computation of the effective inertia matrix $\mathbf{I(\theta)} = \mathbf{D}^{-1}\mathbf{A(\theta)J}^{-1}(\boldsymbol{\theta})$. This would cause the major problem were it not for a very important feature of this application of invariant control.

Returning to eqn. (28) let us suppose that $I(\theta)$ can only be computed approximately so that one may write

$$\mathbf{I(\theta)} = \mathbf{I'(\theta)} + \mathbf{I(\theta)} \tag{30}$$

where $\mathbf{I'(\theta)}$ is the computed approximation to $\mathbf{I}(\theta)$ and $\Delta\mathbf{I(\theta)}$ is the error.

We may now write eqn. (28) as

$$\mathbf{I'(\theta)\ddot{X}} = \mathbf{Q(\theta, \dot{\theta})} - \mathbf{I(\theta)X} + \mathbf{T} \tag{31}$$

or

$$\mathbf{I'(\theta)\ddot{X}} = \mathbf{Q'(\theta, \dot{\theta}, \ddot{X})} + \mathbf{T} \tag{32}$$

where $\mathbf{Q'} \equiv \mathbf{Q} - \Delta\mathbf{I\ddot{X}}$ (33)

Thus we view discrepancies in our knowledge of the inertia matrix as disturbances.

Since eqn. (32) has the same form as eqn. (28) it is reasonable to apply invariant control to eqn. (32). Provided stability is maintained (and it is not possible in the case of this highly non-linear system to give analytic criteria) the resulting system will act as though it actually *had* an effective inertia matrix of $I'(\theta)$ instead of $I(\theta)$.

It has been shown by simulation studies that quite gross approximations $I'(\theta)$ to $I(\theta)$ can be stably tolerated so that piecewise constant approximations to $I(\theta)$ can be tabulated and stored for real-time look-up in relatively short tables.

The main features of invariant robot control may be summarised as follows.

1. It requires no large amount of computational power.
2. It requires relatively little memory.
3. It requires no accurate model and is robust in the face of parameter changes.
4. It can reject completely the effect of outside disturbances.

The method thus offers perhaps the only practical method for fast dynamic control of robots yet proposed.

Much work needs to be done, however, in applying the method to actual robotic machinery.

Summary

1. The control problems of robotics are caused by two main features
   (a) kinematic considerations
   (b) dynamic considerations

2. For slow speed robot movements only (a) is important and can be handled by dedicated computing power.

3. For high speed movements (b) predominates.

4. If trajectory planning (world to joint coordinate transformation ahead of time) is possible the problem of (b) vanishes.

5. If trajectory planning is not possible one feasible method of overcoming the problem of (b) is invariant control.

Chapter 4

Multivariable control

Dr P. M. Taylor

Introduction

It is important to define the requirements of a robotic system and the engineering constraints within which a solution must be sought before studying the robot control problem in detail. Two main types of task can be identified, 'pick and place' and 'continuous path'. The pick and place tasks require fast movement of an object between two points, the trajectory taken can be chosen arbitrarily provided collisions are avoided. Continuous path requires that the arm follows a given trajectory at all times. The system must be cost-effective, fast in operation, highly reliable, robust to sensor or actuator failure, flexible in the sense that it can be quickly modified to perform a different task, and tolerant of uncertainties in its environment and in its own mechanisms. The robot system itself may be intended to be part of a larger computer aided design and manufacture (CADAM) scheme which will result in constraints on the interface between the robot 'manufacturing' part and the CAD facility.

The most crucial of the above constraints are those of speed and cost. Any controller will have to work in real time using currently available hardware, probably microprocessors. Sensors provide the ability to cope with a partially unknown environment. Many sensors, particularly those involving vision provide vast amounts of data, most of which is irrelevant to the current task. It is therefore necessary to process this information to extract only that which is needed for the current task so that the robot controller can invoke the appropriate arm responses. In such circumstances a dedicated sensory data processor is used. A possible scheme for a sensory robot system is shown in Fig. 1.

The overall system controller supervises the task. It interrogates the sensory data processors to obtain information about the environment and as a result instructs the local robot controller to cause the robot arm to make an appropriate movement by sending instructions to the joint servo controller. Feedback from joint sensors is available to the controllers.

Currently available commercial robots make little use of sensory feedback. They rely on high mechanical precision in the construction of the arm and its joint sensors, combined with fairly rigid arms to obtain precise control of the end effector. This makes them very cumbersome when compared with, say, a human arm. It also restricts their applications. They cannot continually monitor the task during continuous path operation or cope with variations in the presentation of objects in pick and place operations.

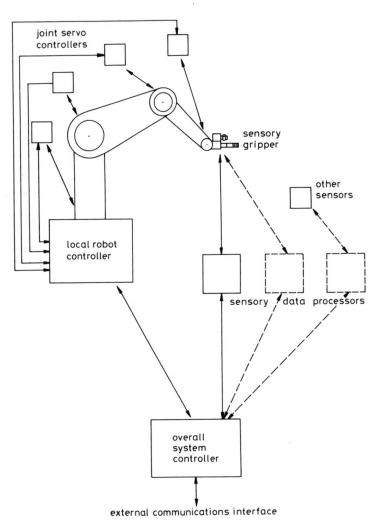

Fig. 1. *Schematic of possible sensory robot system*

Any requirement for full sensory feedback has important repercussions on the design of the control system. If, as a rough guide, the bandwidth of a robot is taken to be about 10 Hz, this implies a sampling rate of say about 50 Hz in order to obtain good control. The local controller must therefore calculate its commands in 20 msec or less unless the bandwidth is artificially reduced. Consequently any control algorithm used must be relatively simple if a microcomputer is to be used. Unfortunately the design of current industrial robots means that they are difficult to control being multivariable, highly nonlinear, and time varying. The next section discusses some approaches to the control problem.

Linear multivariable control

Classical control theory as taught to most undergraduate engineers deals with the control of single input single output time invariant systems which have been modelled by linearising about a chosen operating point. A frequency domain design or root locus design is used to obtain stability and good closed loop time responses. One of the purposes of closed loop feedback is to obtain a degree of insensitivity to disturbances and to modest changes in the system being controlled.

Such techniques have been extended to the multivariable case where the system has several inputs and outputs and extra problems arise because of the interaction occuring between loops. Design methods using the inverse Nyquist array,[1] the characteristic loci[2] and multivariable root loci[3] rely on a linear model of the system. Usually however the designs are robust enough to tolerate small nonlinearities. An alternative to frequency domain analysis is to use state space representation and pole shifting techniques.[4] However, this can lead to less robust designs.

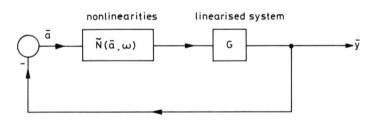

Fig. 2. *Nonlinear feedback control system*

Consider the case of a three degree of freedom robot mechanism as shown in Fig. 2. This system can be adequately described[5] by the state space model

$$\dot{x} = A(x, t) + B(x, t)U(t) \tag{1}$$

$$y(t) = C(x, t) \tag{2}$$

where x is a 6 dimensional state vector, $U(t)$ is a 3 dimensional forcing input vector and $y(t)$ is a 3 dimensional output vector.

A linear asymptotically stable, time invariant model of the desired closed loop system behaviour is postulated as

$$\dot{x}_m = A_m x_m + \lambda R(t) \tag{3}$$

$$y_m = C_m x_m \tag{4}$$

where A_m and λ are scalar matrices and λ is diagonal giving non-interactive control, $R(t)$ is the desired vector.

One approach[6,7,8] is to design a closed loop control system which drives the error ϵ between the desired state x_m and the system state x to zero in a short period.

Thus,

$$\epsilon = x - x_m \tag{5}$$

Subtracting eqn. (3) from eqn. (1) gives

$$(\dot{x} - \dot{x}_m) = A_m(x - x_m) + (A - A_m x) + BU(t) - \lambda R(t) \tag{6}$$

Substituting for x_m from eqn. (5) gives

$$\dot{\epsilon} = A_m \epsilon + (A - A_m x) + BU(t) - \lambda R(t) \tag{7}$$

If we choose the control vector

$$U(t) = (B^T B)^{-1} B^T (A_m x - A) + (B^T B)^{-1} B^T \lambda R(t) \tag{8}$$

then $\dot{\epsilon} = A_m \epsilon \tag{9}$

If the eigenvalues of A_m can be made to lie deep in the left-half plane then $\epsilon(t)$ will decay to zero very quickly.

It can be shown[6] that, using a linearised model of the system shown in Fig. 2., with joint velocities zero, the control vector is given by

$$U(t) = \begin{bmatrix} T_1 \\ T_2 \\ T_3 \end{bmatrix} = -\begin{bmatrix} K_{11} & K_{12} & K_{13} & K_{14} & 0 & 0 \\ 0 & K_{22} & K_{23} & 0 & K_{25} & K_{26} \\ 0 & K_{32} & K_{33} & 0 & K_{35} & K_{36} \end{bmatrix} \begin{bmatrix} \theta_1 - r_1 \\ \theta_2 - r_2 \\ \theta_3 - r_3 \\ \dot{\theta}_1 \\ \dot{\theta}_2 \\ \dot{\theta}_3 \end{bmatrix} \tag{10}$$

where T_i is the torque applied to joint i, r_i is the desired angle of joint i and θ_i is the measured angle of joint i.

The feedback gain matrix K has elements K_{ij} which are constants for the given operating point. Any change in operating point would result in changes in K_{ij}. In particular the model was linearized about the operating point having zero joint velocities. The control system, although satisfactory for slow motion, is unable to force the arm to follow the required trajectory under fast motion. One way around this problem is to have several precomputed K matrices stored in the controlling ccmputer, the one used depending on the current operating point.

Other approaches have been described by Hewit[9] and Wellstead[10] so will not be discussed here.

Nonlinear multivariable control

Many systems, such as robots, are inherently nonlinear. In addition, in many cases, the optional controller to use is nonlinear rather than linear. The main drawback to the analysis and design of nonlinear systems is the difficulty of the mathematical analysis.

Second order single input single output nonlinear systems can be designed using the phase plane method[11] where the relationship between the two states x and $\dot{x}$ can be shown as time varies by a series of trajectories on the $x, \dot{x}$ plane for a series of initial conditions. Analysis and design of a nth order system requires a n dimensional space to be used which makes human visualisation difficult.

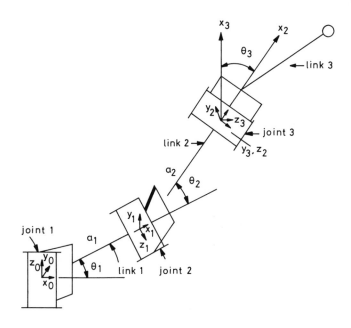

Fig. 3. *Three degree of freedom manipulator*

Several techniques have been developed for analysis and design in the frequency domain. The methods of Popov,[12] and Circle Criteria[13] rely on separable linear and nonlinear elements as shown in Fig. 3. This can occur when a system can be adequately linearised but the control actuators are highly nonlinear. The circle criteria have been extended to multivariable systems[14] but again require the system structure given in Fig. 3.

An alternative method makes use of the describing function approach[15] where a series of linear approximations are made to the nonlinearities for a series of amplitudes of input signals to the nonlinearities. In the traditional use of the method it is assumed that the open loop system is low pass so that all harmonics are completely attenuated. However, refinements to the method have been made, notably by Mees[16] which take into account the presence of harmonics.

Multivariable nonlinear systems were studied by Mees[17] who used a combination of the inverse Nyquist array and its Gershgorin bands with the describing function approach. Gray and Taylor,[18] using describing functions, devised a sequential computational method based on the Nyquist Array which produces much more precise analyses and designs than the Mees method. Displays are produced in the frequency domain, from which the designer decides on an appropriate compensator to avoid limit cycle operation. This method has been used by Gray and Al-Janabi[19] to design an approximate controller which is then used as a starting point in a numerical search procedure, based on Zakian's method of inequalities[20] to find a controller which satisfies specified constraints in the time domain. This work currently requires seperable nonlinear and linear elements but is extendable.

Freund[7,8] and Ali[6] have developed a nonlinear control system for the three degree of freedom robot described in the previous section. They show that in order to obtain the desired dynamic behaviour given by eqns. (3) and (4) the control vector $U(t)$ is given by

$$U(t) = \begin{bmatrix} T_1 \\ T_2 \\ T_3 \end{bmatrix} \tag{11}$$

where

$$T_1 = -\frac{1}{\Delta} b_{41} \Delta_{11} (d_2 \phi_1 + d_1 \phi_4 - \lambda_1 r_1 + a_4) \tag{12}$$

$$T = -\frac{1}{\Delta} \{ (b_{52} \Delta_{22} + b_{53} \Delta_{32})(d_4 \phi_2 + d_3 \phi_5 + a_5 - \lambda_2 r_2)$$

$$+ (\Delta_{22} b_{53} + \Delta_{32} b_{63})(d_6 \phi_3 + d_5 \phi_6 + a_6 - \lambda_3 r_3) \} \tag{13}$$

and $$T_3 = -\frac{1}{\Delta} \{ (b_{52} \Delta_{23} + b_{53} \Delta_{33})(d_4 \phi_2 + d_3 \phi_5 + a_5 - \lambda_2 r_2)$$

$$+ (\Delta_{23} b_{53} + \Delta_{33} b_{63})(d_6 \phi_3 + d_5 \phi_6 + a_6 - \lambda_3 r_3) \} \tag{14}$$

and θ_i = measured joint angles θ_i $\qquad i = 1, 3$

$\qquad \theta_i$ = measured joint angular velocities θ_{i-3} $\qquad i = 4, 6$

$\qquad a_1 = \phi_4, a_2 = \phi_5, a_3 = \phi_6 ; a_4, a_5, a_6,$

are nonlinear functions, the d_i are constants which define the required eigenvalues, the r_i are desired values of θ_i, Δ, Δ_{ij} are nonlinear functions and b_{ij} are nonlinear functions.

The full definitions of each of these variables are given by Ali.[5,6]

The advantage of this control system is that, if implemented, it would not require any adjustment when the operating conditions change and therefore should be given good control over all speeds. The disadvantage is the large amount of computation required although this would be much reduced by the use of appropriate lookup tables. Note also that the model used in the above analysis assumed ideal linear actuators. It would need further complicating in order to take their nonlinearities into account.

Purdue University research work

The group at Purdue University headed by Luh and Paul have done much work on the theory and application of advanced industrial control. Their approach can be summarised as follows:

1. Compute the (sub) optimal trajectory off line and store all desired positions, velocities and accelerations (21).
2. Use 'resolved acceleration control' (22) finding for position control of the manipulator. This requires the correct input torques to apply to the joints for the set of positions, velocities and accelerations. Their requirement was to compute these torques with a frequency of greater than 60 Hz for the Stanford arm carrying a variable or unknown load along a pre-planned path. Velocities and accelerations are successively transformed from the base of the manipulator out to the gripper, link by link. Forces are then transformed back from the gripper to the base to obtain the joint torques.[23] Computation increases linearly with the number of links. Their overall control scheme is shown in Fig. 4. Their programs are written in floating point assembly language, having an average execution time of 11.5 msec on a PDP 11/45 computer (4.5 msec for computation of input torques) giving a sampling frequency of 87 Hz. They are currently working on a multiple CPU implementation of their work and studying the problems of scheduling in parallel computation.[24]

The future

As computer hardware becomes relatively cheaper better control systems can be implemented in a cost effective manner. Sensors, combined with good control software will add flexibility to robot systems. The use of sensory feedback in closed loop control has important implications for robot design. The need for highly precise mechanical structures will disappear for many applications. The important points will be the quality of the sensors and the speed and accuracy of the control software. Good control should lead to cheaper, lighter and therefore faster mechanical structures.

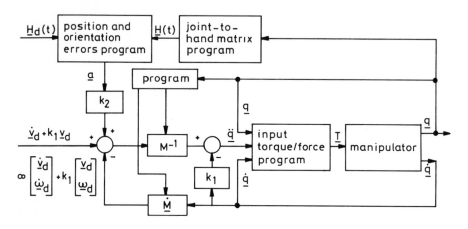

Fig. 4.   *Resolved-acceleration control algorithm*

References

1. ROSENBROCK, H. H. (1974) 'Computer Aided Control Systems Design! Academic Press.
2. MACFARLANE, A. G. J., and KOUVARITAKIS, B. (1977). 'A design technique for linear multivariable feedback systems'. *Int. J. Control,* **25**, pp. 837–874.
3. MACFARLANE, A. G. J., and POSTLETHWAITE, I. (1977). 'Characteristic frequency functions and characteristic gain functions'. *Int. J. Control,* **26**, pp. 265–278.
4. RETALLACK, D. G., and MACFARLANE, A. G. J. (1970) 'Pole shifting techniques for multivariable feedback systems'. *Proc. IEE.* Vol. 117, No. 5. pp. 1037–1038.
5. ALI, A. M. (1981) 'An investigation into the dynamics and kinematics of a manipulator arm'. University of Hull Robotics Research Unit, Control and Software Group, Report No. 1/81.
6. ALI, A. M. (1981) 'Dynamics and control of a manipulator arm' University of Hull Robotics Research Unit, Control and Software Group Report no. 2/81.
7. FREUND, E., and SYRBE, M. (1976) 'Control of industrial robots by means of microprocessors'. Proc. Intn'- Symposium on New Trends in Systems Analysis, Versailles pp. 167–185.
8. FREUND, E. (1977) 'A nonlinear control concept for computer controlled manipulators', Proc. 4th Int. IFAC Symposium MVTS. Fredericton pp. 395–403.
9. HEWIT, J. (1981) 'The robot control problem' SERC Vacation School in Robot Technology, University of Hull.
10. WELLSTEAD, P. E. (1981) 'Intelligent digital control systems' SERC Vacation School in Robot Technology, University of Hull.

11. THALER, G. J. and PASTEL, M. P. (1962) 'Analysis and Design of Nonlinear Feedback Control Systems', McGraw-Hill.

12. AIZERMAN, M. A. and GANTMACHER, F. R. (1964) 'Absolute Stability of Regulator Systems', Holden-Day.

13. ZAMES, G. (1966) 'On the input/output stability of time varying nonlinear feedback systems' *IEEE Trans. on Automatic Control,* AC 11(2) pp 228–238, AC 11(3) pp 465–476.

14. COOK, P. A. (1973) 'Modified multivariable circle theorems' in D. J. Bell (ed) Recent Mathematical Developemnts in Control. Academic Press.

15. ATHERTON, D. P. (1975) Nonlinear Control Engineering, Van Nostrand Reinhold.

16. MEES, A. I. (1972) 'The describing function matrix' *J. Inst. Maths and Appl.* Vol. 10, pp 49–67.

17. MEES, A. I. (1973) 'Describing functions, circle criteria and multiloop feedback systems'. *Proc. IEE* 120(1), p 126–

18. GRAY, J. O. and TAYLOR, P. M. (1979) 'Computer aided design of multivariable non-linear control systems using frequency domain techniques' *Automatica,* Vol. 15 pp 281–297.

19. GRAY, J. O. and AL-JANABI, T. H. (1977) 'The numerical design of multivariable non-linear feedback systems' Proc. 4th Int. IFAC. Symp. MVTS, Fredericton, pp 233–238.

20. ZAKIAN, V. and AL-NAIB, U. (1973) 'Design of dynamical control systems by the method of inequalities' *Proc. IEE* 120, pp 1421–1427.

21. LUH, J. Y. S. and LIN, C. S. (1981) 'Optimal path planning for mechanical manipulators', *Trans ASME* Vol. 103 pp 142–151.

22. LUH, J. Y. S., WALKER, M. W. and PAUL, R. P. C. (1980) 'Resolved – acceleration control of mechanical manipulators', *IEEE Trans. Auto. Control* Vol. AC-25, No. 3 pp 468–474.

23. LUH, J. Y. S., WALKER, M. W. and PAUL, R. P. C. (1980) 'On line computational scheme for mechanical manipulators' *J. Dyanmic Systems, Meas. and Cont*: Vol. 102, pp 69–76.

24. LUH, J. Y. S. (1981) 'Scheduling of distributed computer control systems for industrial robots', *Proc. IFAC,* DCCS Workshop, Beijing, China.

Chapter 5

Control of robot dynamics by microcomputers

P. J. Drazan

1. Introduction

Different aspects of the 'Placemate' robot system developed at Surrey University have been discussed at various conferences therefore only its brief description is presented here. The photograph in Fig. 1 shows the robot designed as an articulated arm consisting of three main degrees of freedom: a base rotating horizontally and two sections, approach and lift, rotating in a vertical plane. Additional degrees of freedom may be added as required, e.g. a gripper, wrist movement, etc. The robot is powered by pneumatic motors which are activated by on-off solenoid valves; it is a teachable machine (point-to-point path control) controlled by a programme called 'SUMOS' (Surrey University Manipulator Operating System), resident in an INTEL 8080 based microcomputer system.

The aim of the lecture is to describe how the computer is used to control the robot. The first part of the paper describes SUMOS as an operating system which consists of TEACH, EDIT and REPLAY modes. The latter part deals more specifically with those sections of the REPLAY mode concerned with the acquisition of the current state of the robot and its dynamic control, the utilisation of the computer time on-line and the future trends in the application of microcomputers to enhance the controllability of robotic machines.

2. Robot operating system

The robot is driven, via a suitable interface, by an Intel 8080-based computer under the overall control of a software operating system known as SUMOS. The primary function of SUMOS is to control the manipulator dynamically while it is driven through a work-cycle previously established in the computer memory. The user first programmes the sequence of path locations, together with certain additional functions, by manually stepping the manipulator along the desired path. At each station in the work-cycle, a key is depressed to instruct SUMOS to store the co-ordinates of that point in the computer memory; this procedure is known as the 'teach' mode. Once the stored sequence is complete, the user may then switch SUMOS into the replay mode when the work-cycle will be reproduced until an instruction is given to arrest the manipulator. A further feature of SUMOS which can be accessed is a mode to enable the user to display and modify, if necessary, any functional element in the taught sequence. This process, known as editing, consists of stepping the manipulator through the work-cycle and, where appropriate, erasing or inserting any desired element.

Fig. 1.

A typical operating environment for SUMOS would consist of an Intel 8080 development system with a minimum of 10 kilobytes of random access memory (RAM), a suitable interface unit to the manipulator sensors and actuators, and a console keyboard and display. SUMOS occupies approximately 7 kilobytes of memory, the remainder being utilised to store data and the user-assigned driving sequence. The interface consists of 8-bit parallel ports which transfer data to monitor the current state of the manipulator, derived from shaft encoders on each degree of freedom, and to provide signals to drive the manipulator through the desired path. A convenient form of console device, which is used currently at Surrey, utilises a hand-held teletype. Because the keyboard is ASCIL-coded, this allows considerable flexibility in re-assigning, through software, the functional significance of each key.

The executive of SUMOS gives user access to any of the facilities available as part of the operating system. A full version of SUMOS consists of three modes:

TEACH, REPLAY and EDIT.

When the system is powered up, all system files are initialised with null arrays and the command level of SUMOS is entered. At this stage the first task is to teach the system a work-cycle before either the replay or editor mode may be entered. In the teach mode functional elements are inserted into a sequential array in memory by depressing appropriate keys at the system console. Representation of these elements are:

(a) Co-ordinate elements, both fine and coarse, the arguments of which are generated automatically by SUMOS.
(b) Input and output interlock control elements, which allow external control of the manipulator during replay.
(c) Subarray access, which allows user to insert, as a separately taught array, a repetitive sequence of co-ordinate elements only differing by small spacial displacements: this facility is useful when a work-cycle includes some form of palletising operation.
(d) Variable time-out element, where a small delay is required in a work-cycle before the next element is accessed and obeyed.

Once all necessary array files have been created by user, SUMOS allows access to all other facilities including replay and editing. When replay is activated, user is requested to place the manipulator in a primary position which is regarded as 'safe' from the operation point of view. Once this condition is satisfied, then the manipulator is automatically locked in position and user may then instruct the system to proceed to the established work-cycle. If necessary, at each of these start-up stages, including the final replay mode, an emergency stop action may be requested by user which depressurises all actuators to produce a manipulator safe state.

During replay, each functional element in the work-cycle array is accessed in turn and satisfied in all respects before the next in sequence is obeyed. The most frequently accessed array element is the co-ordinate element, treated during replay as coarse or fine. SUMOS loads the current co-ordinate into specified locations in memory where this data is treated as a spatial target for the manipulator to drive towards under the action of the control algorithm. In the case of a fine co-ordinate element, the target data must be simultaneously satisfied at full resolution before SUMOS accessed the next sequential control element in the array. When a coarse co-ordinate element is encountered, the stored data is required to be satisfied only to within a specific region surrounding the target; the bit resolution depends on the current version of SUMOS being operated, but is usually ± 3 bits.

Input interlock elements are satisfied when an external signal is set into a specified state, thus allowing equipment outside the immediate manipulator environment to control the progress of the work-cycle. Similarly, output interlock elements allow SUMOS at an appropriate stage in the work-cycle, to activate specific signal lines attached to external equipment or to the manipulator structure, e.g: activation of a gripper, $90°$ rotation of a wrist, etc.

Once SUMOS has been taught a new work-cycle, it is often necessary to modify this in some specific detail in the light of experience. This might be to include an additional sequence of elements, to change an element (especially a co-ordinate element) or even to eradicate an element or elements from the array. In the editor mode of SUMOS, the user may dynamically step the manipulator through its work-cycle, each element in turn being displayed by the hardware. If a particular element requires to be inserted into the array, user steps through to that element immediately following the new one, requests an insertion and presses the appropriate element key. To erase an element, user first diplays that element and then requests an erasure; to change an element, an erasure is first requested, as described above, immediately followed by an insertion. These facilities have proved to be an asset to the SUMOS user and are regarded as an important feature of the operating system.

SUMOS also provides a comprehensive error trapping service; if an illegal proceedure is attempted, a coded diagnostic message is displayed at the console which informs user of his mistakes. For example, an attempt to replay a work-cycle which has not been taught would be trapped and flagged and then control returned to SUMOS command level.

3. Acquisition of data from the robot

In order to control the dynamics of the robot the state of the device must be known at any instance so that the correct control decisions can be taken. The minimal data necessary to control the robot are its current position vector φ and velocity vector ω. To date, the position vector has been obtained from shaft encoders fixed to the individual robot axes. Typically, the Gray coded position read from the encoders is first converted by hardwired decoders to the direct binary codes which are then sampled by the computer via the computer input ports.

The velocity vector ω may be either measured directly by suitable velocity sensors or the velocities may be estimated from the positions. The second approach has been adopted to avoid the need for velocity sensors.

In principle, the velocity $\boldsymbol{\omega}$ may be obtained as the difference between two consecutive positions, i.e.

$$\boldsymbol{\omega} = \frac{1}{\Delta t}\,(\boldsymbol{\varphi}_n - \boldsymbol{\varphi}_{n-1}); \qquad \boldsymbol{\omega} = \frac{\Delta\boldsymbol{\varphi}}{\Delta t}$$

where n is the index of the consecutive readings sampled at intervals.

To improve the estimation accuracy, positions are sampled at a higher rate, e.g. at every millisecond, and the velocity is estimated from the population N as follows:

$$\Delta\boldsymbol{\varphi}_i = \boldsymbol{\varphi}_{N+i} - \boldsymbol{\varphi}_i$$

$$\boldsymbol{\omega}_i = \frac{\Delta\boldsymbol{\varphi}i}{N \cdot \Delta t}$$

$$\hat{\boldsymbol{\omega}} = \frac{\displaystyle\sum_{i=1}^{N} \boldsymbol{\varphi}_{N+i} - \sum_{i=1}^{N} \boldsymbol{\varphi}_i}{N^2 \cdot \Delta t} \qquad (1)$$

An interrupt routine is used to produce estimates of velocity. It is entered every millisecond, causing the position components ϕ_x, ϕ_y and ϕ_z to be read in turn, and the co-ordinate populations accumulated to produce the population sums

$$\left[\sum_{i=1}^{N} \phi_i\right]_{k=x,y,z}$$

On every Nth sample the two latest populations are subtracted to generate the difference

$$\sum_{i=1}^{N} \boldsymbol{\varphi}_{N+i} - \sum_{i=1}^{N} \boldsymbol{\varphi}_i$$

This resulting difference is then used by the control algorithm as a scaled version of the velocity estimation. The second difference may be obtained in a similar manner to estimate acceleration ϵ.

4. Identification and modelling of the robot system

One of the most important aspects of the work was to use the microcomputer 'on-line' to control the robot dynamics. Since 1972 our interest has been to use computers for the Direct Digital Control (DDC) and the robot provided a good testing base for such an approach. A computer used as a programmable controller offers the flexibility which could not be provided by conventional controllers.

The flexibility was even more desirable in the case of Placemate with its highly non-linear characteristics due to using pneumatic motors and on-off valves. More powerful processing of data by the computer enhances the observability of the system thus reducing the number of directly measured parameters.

The modern control approach would not rely blindly on the feedback but would strive to identify the controlled system in order to match correctly the controller. The case of the Placemate is used to illustrate such an approach.

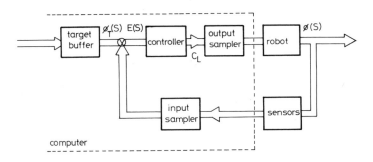

Fig. 2.

The control loop of the robot is shown in Fig. 2. The position of the sections is monitored by encoders and compared with the position currently held in the computer buffer. Any difference produces an error which is processed by a controlling compensating algorithm resident in the computer. The sign of the resultant control signal is sampled out to actuate a pair of on-off solenoid valves each of them controlling one side of a pneumatic motor, thus moving the section.

Let us describe briefly two main components contained in the robot servo-loop, i.e. the valve and the motor. The valve is of a 3 port-2 way type and its schematic is shown in Fig. 3. It has an inlet port 1, an outlet port 2 and vent 3. The plunger 4, moved by energising the solenoid 5, controls the opening of the nozzle 6. By lifting the plunger from the nozzle the output is connected to the air supply and the vent is closed by the other face of the plunger. When the plunger is pressed against the nozzle thus cutting off the supply, the air from the output side is exhausted through the vent.

The pneumatic motor is realised by either using a pneumatic cylinder or a rotary motor of a vane type. A typical representative of the vane motor is one made by KINETROL which uses a single vane rotating through approximately 90°. When a cylinder is used for rotary motion the linear movement would be converted into rotation by using a rack and pinion gear. Torque convertors are readily available from the manufacturers of pneumatic equipment.

The function of the valve-motor pair is determined by the flow characteristics of the valves used and the variable air volumes of the motor. The flow rate of mass through the valve is primarily the function of its upstream and downstream pressures. The inlet supply port of the valve is connected to a source of constant pressure P_0, usually 6 bars. The valve outlet is connected to one side of the motor and the vent to the ambient atmospheric pressure P_A.

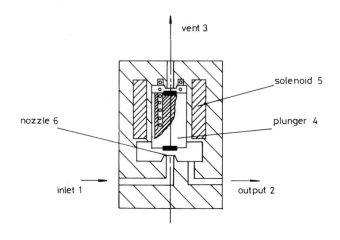

Fig. 3.

One method of determining the valve flow characteristics is as follows:

The pressure transients for both the pressure rise and decay are measured for the valve outlet connected to a fixed volume.

By differentiating the gas equation

$$P \cdot V = M \cdot R_G \cdot T \tag{2}$$

where P = absolute pressure, V = fixed volume, M = total mass of air, R_G = gas constant and T = absolute temperature.

For constant volume and constant temperature we obtain

$$\dot{M} = \frac{V}{R_G \cdot T} \cdot \frac{dP}{dt} \tag{3}$$

which relates the mass flow rate of air passing through the valve to the gradient of the pressure transient. By linking together these flow rates with the corresponding values of the pressure the pair of flow characteristics shown in Fig. 4 can be obtained. The flow characteristics identify the valve as a component contained in the servo-loop.

The next component to consider is the motor itself. Each side of the motor is controlled by one valve. Under normal circumstances one side of the motor is always pressurised whilst the other side is vented. The regime of both motor volumes

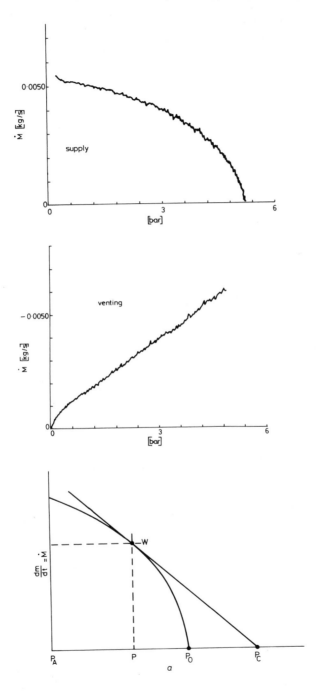

Fig. 4.

are governed by the same gas eqn. (1). When the equation is differentiated in respect to both the pressure and the volume it becomes

$$dM = \frac{1}{R_G T}\left(\frac{\partial M}{\partial V}dV + \frac{\partial M}{\partial P}dP\right)$$

$$dM = \frac{1}{R_G T}(PdV + VdP)$$

$$\frac{dM}{M} = \frac{dV}{V} + \frac{dP}{P} \tag{4a}$$

The volume V is proportional to the motor displacement x, $V = A \cdot x$, where A is the area of the motor piston or vane. It is evident from equation (3) that the pressure in each side of the motor results from the mass of air injected into the respective motor volume (or vented from) and the motor displacement, thus

$$\frac{dP}{P} = \frac{dM}{M} - \frac{d\phi}{\phi} \tag{4b}$$

The torque delivered by the motor to the robot section results from the difference between the two pressures in the two sides of the motor.

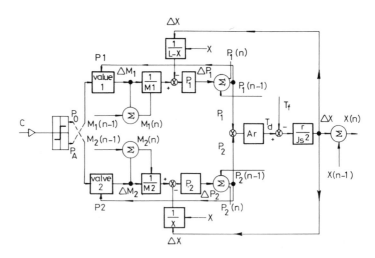

Fig. 5.

A numerical dynamic model of the robot section has been developed using the relationships explained above. The block diagram illustrating the model is shown in Fig. 5. The model is used to simulate the dynamic responses on the computer and to test different control algorithms. The diagram indicates two major 'physical' feedbacks present in the model. The first is the motor pressure line brought back to the valve to compute the flow rates. The second is the line correcting the respective motor volumes due to the change of the robot section position.

To understand the character of the system and to synthesise it with a controller, an attempt is made to linearise the system by considering only small perturbations. Let us represent the middle section of the flow characteristic shown in Fig. 4a by a straight line

$$\frac{dM}{dt} = \frac{P_c - P}{R} \tag{5}$$

where R = valve resistance and P_c = virtual supply pressure.

When the gas eqn. (3) is substituted into the equation above, the pressure changes in the motor are governed by the first order lag relationship

$$\left(1 - \frac{V(\phi) \cdot R}{R_G \cdot T} s\right) P = P_c$$

$$\frac{P}{P_c} = \frac{1}{1 + \tau s} \tag{6}$$

where τ is the time constant calculated from the volume of the motor corresponding to the current position ϕ, from the impedance of the valve R, the gas constant R_G and the absolute temperature of air T.

As seen from Fig. 5, the pressures P_1 and P_2 on the respective sides of the motor determine the driving torque T_d thus leading to the overall transfer function for the robot section

$$\frac{\Phi(s)}{E(s)} = \frac{T_M}{J} \frac{1}{s^2} \left(\frac{1}{1 + \tau_1 s} + \frac{1}{1 + \tau_2 s}\right) \tag{7}$$

where ϕ = displacement of the section, e = system error, τ_1, τ_2 = time constants of the respective sides of the motor and T_M = maximum driving torque corresponding to the maximum pressure difference $P_0 - P_A$ or, in a re-arranged form,

$$\frac{\Phi(s)}{E(s)} = \frac{2T_M}{J} \frac{1}{s^2} \frac{1 + \dfrac{\tau_m}{2} s}{\tau^2(\phi)s + \tau_m s + 1} \tag{8}$$

where $\tau_m = \tau_1 + \tau_2$ is a constant and $\tau^2(\phi) = \tau_1 \tau_2$ is a coefficient which value is dependent on the section position.

A root-locus of the system in Fig. 6 corresponding to the median position shows that the system is inherently unstable.

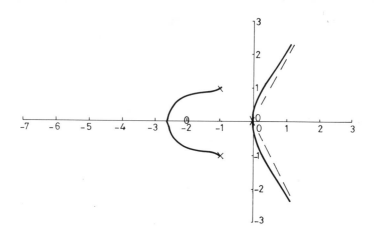

Fig. 6.

5. Control of robot dynamics

The control loop of the robot was shown in Fig. 2. The target vector φ_T is compared with the current position vector φ sampled from the encoders. The control vector processed by the computer is sent out via the output computer ports to energise the valves actuating the robot motors which, in turn, power the robot sections. The various sections of the robot, i.e. rotation, approach and lift, are considered fully decoupled and are serviced by the computer sequentially, using a common control algorithm with control parameters set to match the dynamics of the individual sections.

The control algorithm used to data is a lead term vector

$$\mathbf{C}(s) = (1 + \alpha s + \beta s^2) \cdot E(s) \tag{9}$$

where $E(s) = \Phi_T(s) - \Phi(s)$ is the error vector and α, β are the weighing coefficients of velocity and acceleration respectively.

However, using the on-off valves to actuate the robot, the control vector $\mathbf{C}$ is converted into a set of binary signals by generating the signs of this load components as

$$(C_2)_k = \operatorname{sgn} c_k = \operatorname{sgn} (e_k + \alpha_k \omega_k + \beta_k \epsilon_k)$$

where $k = x, y$ and z represent the individual degrees of freedom of the robot.

The algorithm computation undertaken on-line utilises the AM-9511 arithmetic processor unit on board of an AMC microcomputer. The root-locus of the robot section compensated by a single zero ($\beta = 0$), i.e. by a pure velocity feedback, is shown in Fig. 7. It can be seen that parts of the locus are still unstable demonstrating

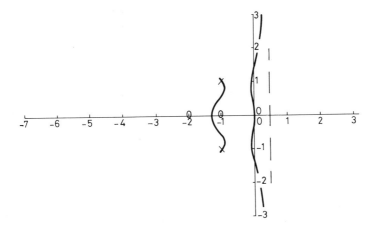

Fig. 7.

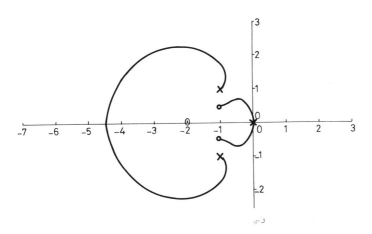

Fig. 8.

that the velocity feedback is inadequate to compensate the type of system in question. A much more satisfactory result is obtained when the second controller form ($\beta \neq 0$) is utilised as well. This result is shown in Figure 8.

6. Time-sharing in the replay mode

The amount of software which can be processed by the computer on-line is determined by the frequency at which the system has to be controlled, in other words, by the output sampling rate. The output sampling rate is determined in turn by the cut-off frequency of the system. The output sampling frequency f_{out} used on the robot in question was chosen to be 50 Hz, i.e. an up-date of the control commands is made every 20 ms.

The available time interval of 20 ms is shared by the interrupt service routine and by the replay section of the programme which includes the control algorithm. A few criteria have to be observed to obtain a satisfactory time sharing arrangement. First, the individual interrupt routines and, in particular, the Nth routine, which takes more time due to completing the velocity and acceleration estimation, must not exceed the input sampling interval $t_{int} = 1/f_{int}$, in the current case 1 millisecond. Second, as seen from Fig. 9, all N interrupts which are part of one estimation, must

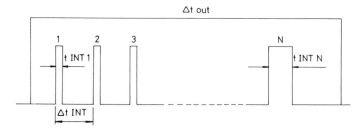

Fig. 9.

be accomplished within one output sampling interval $t_{out} = 1/f_{out}$, i.e. 20 milliseconds, to ensure that an up-dated estimation of velocity is available for the next control interval. This criteria determines the interrupt frequency f_{int} as

$$f_{int} > N \cdot f_{out}$$

Third, sufficient time must be allowed to process the control algorithm as indicated in the following relationship

$$t_{cmax} + \sum_{i=1}^{N} t_{int} < \Delta t_{out}$$

where t_{cmax} is the maximum total time required to process the control loop.

7. Future trends

An accurate mathematical model of the dynamics of the robot was developed to aid its design. So far, it has been used mainly to research optimal control strategies for the robot. It is the aim now to produce a simplified model which could make the system more observable.

Another aspect of the application of microprocessors to robots is to use distributed systems. Primarily the acquisition of data, such as the velocity or acceleration estimation would be undertaken outside the master computer, either by hardwired circuitry or dedicated processors. Furthermore, each degree of freedom, if necessary, could be served by one processor and only the interacting parameters would be transmitted between the sections. Additional functions, such as balances of the dead-weights inherent in the mechanical structure of the robot could then be catered for by selected processors as more processing time would become available.

References

DRAZAN, P. J. and JEFFREY, M. F. 'Some aspects of an Electro-pneumatic Industrial Manipulator' 8th International Symposium on Industrial Robots; Stuttgart, May 1978

DRAZAN, P. J. and ZAREK, J. M. 'Controllability of an Electro-pneumatic Manipulator System' Third CISM-IFToMM on Theory and Practice of Robots and Manipulators; Italy, Udine, September 1978

DRAZAN, P. J. and THOMAS, P. B. 'Simulation of an Electro-pneumatic Manipulator System' 5th International Fluid Power Symposium; Durham, September 1978

DRAZAN, P. J. and JEFFREY, M. F. 'Control of Robot Dynamics by Microcomputers' Microprocessor Workshop, University of Liverpool; Liverpool, September 1979

DRAZAN, P. J. and TOULIS, V. 'The use of interpolation routines for the generation of Robot path' 10th ISIR; Milan, March 1980

DRAZAN, P. J. 'Control of Robot Dynamics' Conf. of Swiss Ass. for Automation; Zurich, March 1980

DRAZAN, P. J. & others. Lecture Notes on Robot Systems; Short Course, University of Surrey, Mechanical Engineering Department, June 1981

Chapter 6

The automatic guidance and control of an unmanned submersible

G. T. Russell

1. Introduction

The remotely controlled unmanned submersible is now a well established offshore engineering tool, providing the basic functions of, (i) data acquisition, (ii) surveillance and inspection, and (iii) remote working.

The submersible is simply a means of extending the faculties of man into an alien environment, giving man the ability to inspect and analyse sub-sea structures or the sea bed; providing a workhorse to aid divers and allowing the remote manipulation of tools. The submersible and associated launch and retrieval equipment is a complex engineering system comprising many component sub-systems, each serving a specific function, working toward achieving the mission objective.

The automatic guidance and control problem becomes one of manoeuvring the submersible mainframe into a location to execute a given task, whether this be in tracking a pipeline or inspecting the nodes of an offshore platform. The solution takes on many forms, but generally the control tasks can be partitioned into a number of levels of given priority within a hierarchical structure. Positional accuracy demands specific instrumentation, inspection requires appropriate video, acoustic or electromagnetic equipment and a work function demands a manipulative arm or tooling device.

2. The guidance and control strategy

There are many alternative strategies for a submersible vehicle, the type and precision of control depending ultimately upon the prescribed mission. In designing the control system there are a number of fundamental stages to be satisfied.

2.1 The mission objective must be identified.

This defines the basic payload, the devices used to collect information and allow the operator to perform the proposed task. The payload and mission, in turn, define the size and shape of the mainframe, e.g. a localised inspection mission may require a small manoeuvrable vehicle that can operate from a tethered cage, whereas a vehicle designed to attach itself to, and clean parts of, an oil rig may be very large, but still require considerable agility. On the other hand, a long range surveillance vehicle may possess an efficient, streamlined shape to give speed in one direction. Fig. 1 shows a selected group of typical submersibles classified in terms of their

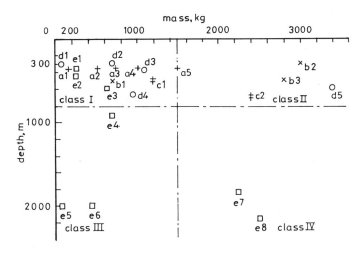

Fig. 1.

mass and operating depth. They fall into four broad classes, where class I vehicles tend to be surveillance and inspection machines for work on or around equipment in the continental shelf regions, whereas class II machines are designed for heavy duty remote working in the same shallow water areas. Similarly, class III machines are primarily for surveillance and inspection and class IV for remote working in the deeper ocean environment. A typical payload would consist of a hydraulically powered tool assembly, perhaps a 5 or 7 function manipulator. Search and navigation equipment could include a compass, a continuous transmission active sonar with a range of 500 m and scanning sector of 30 to 360 degrees, a transponder mode with a 800 m range, an echo sounder for determining the distance to the sea bed, a temperature sensor and a pressure transducer. Underwater viewing could be obtained by a number of Vidicon T. V. Cameras, slow scan CCD cameras or low light cameras giving a viewing range of 1 to 10 m, depending upon the relative scattering, spreading and absorption of the illumination. This could be provided by an array of thalium iodide headlights and mercury-vapour spotlights. Cine or 35 mm still cameras with strobe lights can also be included together with pan and tilt mechanisms controlled from the surface. Each vehicle would be supported by a control station on the surface support vessel, which would house T.V. display monitors, data display monitors for vehicle depth, heading, camera pan and tilt angles, number of cable twists, tether length, inertial heading, position, temperature and time, etc.

2.2 The vehicle must be controllable.

To achieve the desired manoeuvrability the mainframe must have sufficient thrust to operate at the designed speed, or to stem a defined current. Then the thrusters must be positioned such that control can be exerted over all six-degrees-of-freedom. A typical thrust vector configuration can be defined by,

$$R_L = \beta i + \gamma j + \alpha k$$

$$R_\omega = Oi + Oj + \delta k$$

where R_L is the lateral component vector and R_ω is the rotational component vector. The roll and pitch components could be controlled by operating the vertical thrust α, in a directional mode, but this is normally not required since the large buoyancy moment gives the vehicle static stability.

The submersible designed and constructed at Heriot-Watt University, ANGUS 002, is driven by two sets of propellers to give vertical and horizontal thrust. These are coupled to 3-phase induction motors mounted in base pods, thus placing the heavier components well below the centre of buoyancy. Three sets of thyristor inverters provide 3-phase variable frequency supply to the induction motors. The cable is 600 m in length having multiple power conductors, two bundles of screened twisted pairs and two low loss coaxial cables. It is strengthened by a fine high tensile stainless steel braid. The dynamic response of the propulsion system has been determined and typical frequency response characteristics are given.[2] The performance of both vertical and horizontal thrusters varies slightly with forward speed, cross flow current and operation of other thrusters. One of the main inter-actions is between the forward or reverse force induced by the vertical thrusters. Typical characteristics have been derived showing, for example, that an additional normal drag force of approximately 120 N is induced at a forward speed of 0.5 m/s when the vertical thrusters are in operation. This is substantially greater than the drag of the vehicle with all motors off. Consideration of the inflow and outlet conditions to each of the four ducts containing the vertical thrusters gives an appreciation of the mechanism giving rise to this drag force.

The geometry of ANGUS 002 is such that it is extremely stable in pitch and roll, whilst being propelled by thrusters in the horizontal and vertical planes. The inter-active forces and drags can be characterised by preliminary mechanics to describe the vehicle motion by a pair of differential equations.[3]

$$m\dot{V} = F(u, v, w, r, q, p) - (m\omega) \times V$$

$$I\dot{\omega} = H(u, v, w, r, q, p) - \omega \times (I\omega)$$

where

$$V = ui + vj + wk$$

$$\omega = pi + qj + rk$$

The particular motions of interest for the submersible are those in the horizontal and vertical planes and the equations can be conveniently partitioned to represent those components of motion. The incomplete set of inputs in the thrust vector means that the manoeuvrability of the vehicle is restricted and the ability to offset disturbances is limited.

2.3 The primary state variables of the vehicle must be measured.

If the vehicle is to be positioned accurately with respect to a fixed platform or support vessel, as shown in Fig. 2, in the presence of disturbances due to sea current and cable forces, the motion in each of the six-degrees-of-freedom must be observed. Ideally, to hold the vehicle in a hover mode, the total disturbance vector must be measured in all degree of freedom, and an equal and opposite thrust vector must be applied to eliminate the disturbance. Obviously, instrumentation is available to give the total disturbance vector, but this is expensive and some form of compromise is usually sought. The ANGUS 002 vehicle has instrumentation to measure the heading and height above the sea bed. A short base line acoustic navigation system is used to measure the vehicle position with respect to the ship or a fixed transponder.

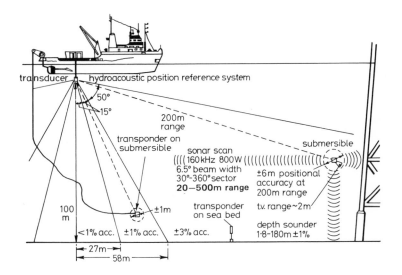

Fig. 2.

2.4 The implementation of the control strategy.

For the particular applications envisaged for the ANGUS 002 vehicle a surveillance and inspection mission was identified. The general control strategy was structured in a number of levels as shown in Fig. 3, forming a hierarchy of control. Manual control of the thruster demand is the basic level 0, then with minimum instrumentation on the vehicle the heading and height/depth can be measured. This forms the level 1 computer control, in which heading and height can be specified from the computer keyboard and trimmed by the pilot using the joystick on the control console. The level 2 loop co-ordinates the geographical range and bearing measurements from the navigation system and estimates the position vector.

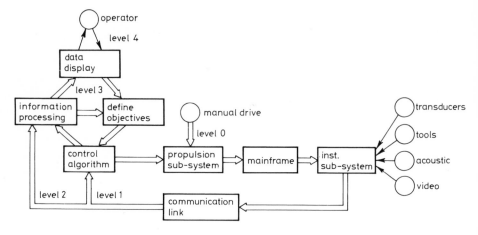

Fig. 3.

Because of the limited thrust vector, the position has to be maintained by controlling the forward thrust and rotation in the $X-Y$ plane, and the vertical thrust in the Z-plane. The position control loop must contain two orders of integration or a feed forward (prediction) characteristic to reduce tracking errors to a minimum. This constraint, together with sea path delay, sea current disturbances and variance in the navigational measurements, puts severe conditions on the stability requirements of the level 2 loop. The man/machine interaction is contained in the level 3 loop, giving the pilot a dynamic status and a three dimensional graphic display.[3] This shows the actual vehicle position or simulated position with a record of the past trajectory over a defined time period. Level 4 of the control structure provides a data log facility to a floppy disc within the control microcomputer, alarm detection, and assessment of the vehicle dynamic performance.

The computer hardware configuration is shown in Fig. 4. The operator is presented with three data displays in addition to a normal T.V. monitor. A 3-dimensional image of the vehicle and past trajectory, with respect to the ship is given on the graphics terminal. A typical display is shown in Fig. 5.

The software is structured within a real-time multi-task, disc operating system running on the D.E.C. L.S.I 11/02 computer.[5] The real-time operation schedules the software tasks and two links to remote hardware tasks. Firstly, the Motorola M6800 microprocessor on the vehicle, acting as an intelligent multiplex for the instrumentation data, and secondly, the SIMRAD acoustic navigation system. The software tasks consist of the control levels 1, 2, 3 and 4 respectively.

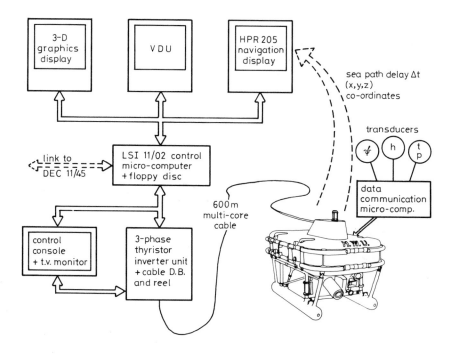

Fig. 4.

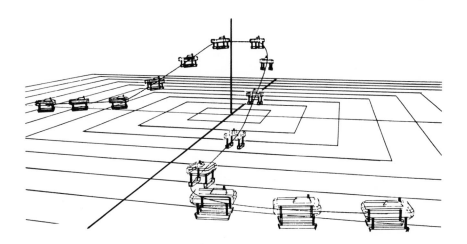

Fig. 5.

3. The positional control algorithms

Two distinctive control algorithms are employed in level 1 and level 2 of the strategy as shown in Fig. 6. Conventional sampled-data techniques are used in level 1 to achieve a near dead-beat response for heading and height/depth. The guidance control loop utilises a prediction-correction estimator to minimise the effect of measurement noise.

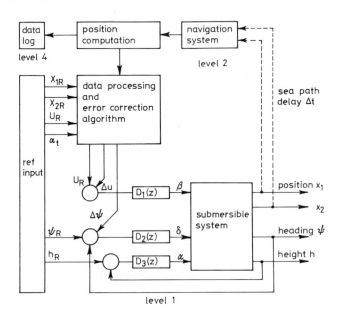

Fig. 6.

3.1 Level 1 control loop.[4]

For normal reference changes in heading and height, the interaction between these variables can be considered to be small. Both loops can take the same form as that shown in Fig. 7. For this particular loop the transfer function relating vertical velocity to input thrust can be given to a first approximation by the expression,

$$M_t \dot{W}(t) + K_d W(t)|W(t)| = \alpha(t)$$

where M_t is the combined mass of the vehicle and the added mass of the entrained water, and K_d is the drag coefficient.

Within the control software two digital filters $D_1(nT)$ and $D_2(nT)$, are employed to estimate the rate of change of velocity $\hat{W}(nT)$ and smooth the measured height $z(nT)$, respectively. It has been found that an averaging filter having the discrete transfer function

$$D_1(z) = \frac{1}{3T_s} \frac{(z-1)(z^2+z+1)}{z^3}$$

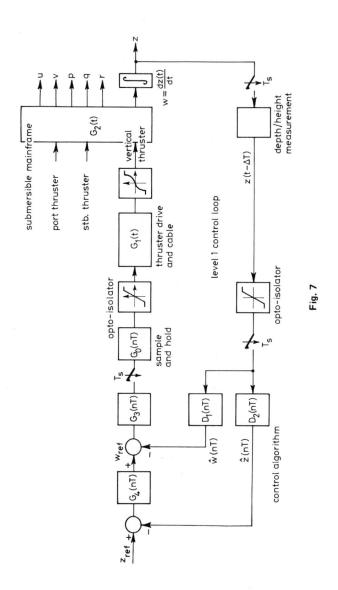

Fig. 7

gives a good estimate of the velocity rate when the sampling time T_s is 100 ms. Similarly, a simple second order low pass filter of the form,

$$D_2(z) = \frac{ze^{-aT}\sin\omega_0 T}{(z^2 - z2e^{-aT}\cos\omega_0 T + e^{-2aT})}$$

provides adequate smoothing for the measured data. The inner loop forms a 'rate' control and the serial compensation element $G_3(nT)$ should have a proportional plus integral characteristic of the form,

$$G_3(z) = K_{prop} + K_{int}T_s(1 - z^{-1})$$

The outer loop is closed through a saturating gain element $G_4(nT)$ so that a defined rate of change is achieved for large changes in the reference input.

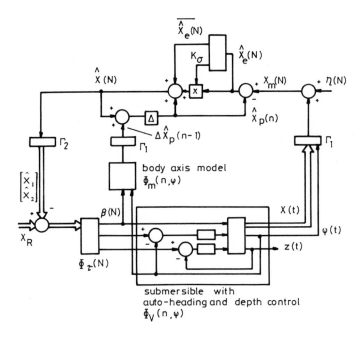

Fig. 8

3.2 *Level 2 guidance loop.*
The first level of control around the body states of the vehicle reduces the effect of the non-linearities so that a simple model can be used to predict the forward and lateral velocities, as shown in Fig. 8, that is

$$\Delta X_p(n-1) = \Gamma_1 \Phi_m(n, \phi) \cdot \beta(N)$$

where Γ_1 is the transformation from body axes to geographical axes, $\Phi_m(n, \phi)$ is the transition matrix of the vehicle model, and $\beta(N)$ is the thrust vector. The estimated position has three components, given by the expression,

$$\hat{X}(N) = \hat{X}_p(N) + K_\sigma [X_m(N) - X_p(N)] + \overline{\hat{X}_e(N)}$$

where the first term $\hat{X}_p(N)$ is the predicted position,

$$\hat{X}_p(n-1) = \Delta \hat{X}_p(n-1) + \hat{X}(N)$$

the second term is the weighted position correction and the third term is the bias correction. The weighting factor K_σ is successively adapted to give minimum estimator error $X_e(N)$, based upon the stochastic properties of the difference between the measurement and the predicted position,

$$\hat{X}_e(N) = X_m(N) - \hat{X}_p(n)$$

The prediction is computed at a constant sample rate n and the measurement is presented asynchronously at a much lower rate N. The best estimate of position can be transformed into longitudinal and transverse components with respect to the reference path, giving the positional error to derive the driving vector $\beta(N)$, that is,

$$\beta(N) = \Phi_c(N)[X_R - \Gamma_2 \hat{X}(N)]$$

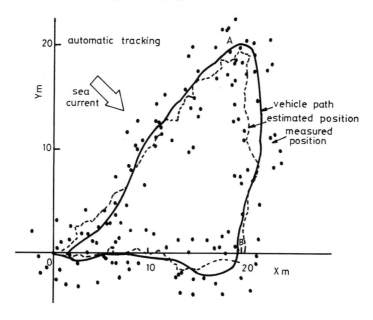

Fig. 9.

The series controller $\Phi_c(N)$ is designed to give the minimum transverse offset, with a constant longitudinal velocity. The detailed analysis of this adapted guidance algorithm is contained in the reference,[6] but a typical result is shown in Fig. 9.

Here the vehicle is programmed to follow the path OA, AB, BO at a velocity of 0.5 m/s, in an equivalent sea current of 30% maximum thrust, with measurement noise of ± 1 m. The vehicle trajectory is adequate even in this extreme operating condition.

4. Acknowledgement

The submersible research programme in the Department of Electrical and Electronic Engineering at Heriot-Watt University has been funded by the SRC Marine Directorate and the Department of Energy.

5. References

1   Janes' Ocean Technology 1979–80, Fourth Edition, L. Trillo.
2   FYFE, A. J. and RUSSELL, G. T., "Closed loop control systems and hydrodynamics of an unmanned submersible", O.T.C. Paper No. 3767, Houston, Texas, USA, May 1980.
3   BELLEC, P., "Simulation of the six-degree-of-freedom motion of a remotely controlled unmanned subermersible", M.Sc. Thesis, Heriot-Watt University, 1980.
4   PANTIGNY, P., "Computer control for the height and depth of an unmanned submersible", M.Sc. Thesis, Heriot-Watt University, 1979.
5   RUSSELL, G. T. and BUGGE, J., "An integrated guidance and control system for a tethered submersible", IEE Conf. on control and its applications, Publication no. 194.
6   BUGGE, J., "A real-time hierarchical computer structure for the design and simulation of automatic control systems", Ph.D. Thesis, Heriot-Watt University (to be submitted 1981).

Chapter 7

Processing of binary images

A. Pugh

Preface

The following notes have been reproduced from Chapters 7 and 9 of The Application of Visual Feedback to Assembly Machines by P. W. Kitchin, Ph.D Thesis, University of Nottingham, October 1977. Dr. Kitchin completed his work over the period 1969–73 and many of the techniques researched at that time have relevance today in the commercial vision systems being marketed. The notes provide for the reader a unique opportunity to study a clearly presented introduction to binary image processing procedures. The notes are by no means comprehensive and represent the research studies of a single research group.

The lecturer gratefully acknowledges the permission of Dr. Kitchin and the University of Nottingham to reproduce parts of this thesis in these notes.

1. Introduction

A considerable amount of research has been reported into methods of processing two dimensional images stored as binary matrices (for bibliographies see Ullman,[1] Rosenfeld[2]). A large part of this work has been directed towards solving problems of character recognition. Whilst many of these techniques are potentially useful in the present context, it is valuable to note some important differences between the requirements of character recognition and those associated with visual feedback for mechanical assembly.

(i) Shape and size
All objects presented to the assembly machine are assumed to be exact templates of the reference object. The objects may be any arbitrary geometric shape, and the number of possible different objects is essentially unlimited. Any deviation in shape or size, allowing for errors introduced by the visual imput system, is a ground for rejection of the object (though this does not imply the intention to perform 100% inspection of components). The derived description must therefore contain all the

shape and size information originally present in the stored image. A character recognition system has in general to tolerate considerable distortion ('style') in the characters to be recognised, the most extreme example being the case of hand-written characters. The basic set of characters is, however, limited. The closest approach to a template matching situation is achieved with the use of type fonts specially designed for optical reading, such as OCR A and OCR B.

(ii) Position and orientation

A component may be presented to the assembly machine in any orientation and any position in the field of view. Though a position and orientation invariant description is required in order to recognise the component, the measurement of these parameters is also an important function of the visual system to enable subsequent manipulation.

While lines of characters may be skewed or bowed, individual characters are normally presented to the reconition system in a relatively constrained orientation, a measurement of which is not required.

(iii) Multiple objects

It is a natural requirement that the visual system for an assembly machine should be able to accommodate a number of components randomly positioned in the field of view. The corresponding problem of segmentation in character recognition is eased (for printed characters) by a priori knowledge of character size and pitch. Such information has enabled techniques for the segmentation of touching characters.[1] No attempt is made in this study to distinguish between touching objects. Their combined image will be treated by the identification procedures as that of a single, supposedly unknown object.

The essentially unlimited size of the set of objects that must be accommodated by the recognition system demands that a detailed description of shape be extracted for each image. There are, however, a number of basic parameters which may be derived from an arbitrary shape to provide valuable classification and position information.

These include

> Area
> Perimeter
> Minimum enclosing rectangle
> Centre of Area
> Minimum radius vector (length and direction)
> Maximum radius vector (length and direction)
> Holes (number, size, position)

Measurements of area and perimeter provide simple classification criteria which are both position and orientation invariant. The dimensionless 'shape' factor (Area)/(Perimeter)^2 has been used as a parameter in objection recognition.[3] The coordinates of the minimum enclosing rectangle provide some information about the size and shape of the object, but this information is orientation dependent. The centre of area is a point that may be readily determined for any object, independent of orientation, and is thus of considerable importance for recognition and location purposes. It provides the origin for the radius vector, defined here as a line from the centre of area to a point on the edge of an object. The radius vectors of maximum and minimum length are potentially useful parameters for determining both identification and orientation. Holes are common features of engineering components, and the number (if any) present in a part is a further suitable parameter.

The holes themselves may also be treated as objects, having shape, size and position relative to the object in which they are found.

The requirements for the establishment of connectivity in the image and the derivation of detailed descriptions of arbitrary geometric shapes are most appropriately met by an edge-following technique. The technique starts with the location of an arbitrary point on the black/white edge of an object in the image (usually by a raster scan). An algorithm is then applied which locates successive connected points on the edge until the complete circumference has been traced and the starting point is reached. If the direction of each edge point traced relative to the previous point is recorded, a one-dimensional description of the object is built up which contains all the information present in the original shape. Such chains of directions have been extensively studied by Freeman.[4-8] Measurements of area, perimeter, centre of area and enclosing rectangle may be produced while the edge is being traced, and the resulting edge description is in a form convenient for the calculation of radius vectors.

Edge-following establishes connectivity for the object being traced. Continuing the raster scan in search of further objects in the stored image then presents the problem of the 're-discovery' of the already traced edge.

Rosenfeld[2] suggests changing all edge points visited by the edge-following routine from '1' to '2's for the first object trace, to '3's for the second object, and so on (where a '1' represents an object, or 'black' in the original image, and '0' represents the background or 'white'). This approach, however, necessitates the provision of more than one bit of storage for each picture point in the image and is consequently not appropriate to a system in which minimising memory usage is important. The solution adopted by the author is to apply a spatially differentiating or edge extracting operator to the image immediately after input from the camera. This operator replaced all '1's in the image by '0's unless they are deemed to lie on a black/white border. A computer plot of the contents of FRAME with the camera viewing a square and a disc is shown in Figure 1, and the result of applying the edge-extracting operator is shown in Figure 2. The edge-following procedure, called EDGETRACE, may now be applied to the image in the same way as for 'solid' objects. The procedure is arranged, however, to reset each edge point as it is traced. The tracing of a complete object thus removes it from the image, and ensures that it will not be subsequently retraced.

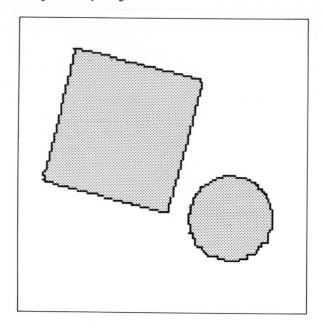

Fig. 1.   *The frame after procedure input-frame*

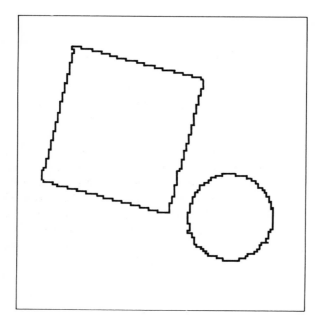

Fig. 2.   *The frame of Figure 1 after edge extraction*

2. Smoothing

Picture processing techniques based on edge-following are likely to fail if the image contrast boundaries are broken. Such operations in character recognition systems are consequently often preceded by a smoothing or local averaging process, which tends to eliminate voids (white elements that should be black) and spurious black points, and to bridge small gaps. In the smoothing technique due to Dineen[9] an $n \times n$ element window is positioned over each element in the image in turn. The total number of black elements in the window are counted. A new image is formed in which each element corresponds to one position of the window, and each element is black only if the number of black elements in the corresponding window exceeds a preset threshold. The smallest practicable window size is 3×3 elements. Unger[10] has proposed a smooth process in which, instead of averaging, logical rules are applied to the contents of a 3×3 window, in order to determine the value of the corresponding element in a new pattern. The disadvantage of both systems in the present context is the relatively large amount of computer bit manipulation involved. Each requires of the order of 100 computer instructions to create a new pattern element.

The objects viewed by the assembly machine, however, are fundamentally different from printed characters, in that any gaps or voids in the objects themselves represent defects, and would be valid grounds for non-recognition. Isolated noise points occur infrequently, and will not in general affect the edge-following algorithm. Consequently, though steps are taken in the following procedures to minimise the effect of noise points, no specific smoothing operation is performed.

3. Edge extraction – Procedure OUTLINEFRAME

The edge extraction operator must produce an unbroken sequence of edge points to ensure the success of the subsequent edge-following program. To minimise computing time, the generation of each element for the new FRAME must involve accessing the smallest possible number of picture elements in the original FRAME.

Rosenfeld[11] distinguishes between the 'border' and the 'edge' of a pattern as shown in Figure 3, defining the border as being made up of the outermost elements of the pattern, and the edge as lying midway between horizontally or vertically adjacent pairs of pattern/background points. The concept of 'edge' rather than 'border' has been adopted for this application as it more accurately represents the true boundary between black and white in the physical world.

It also provides an unambiguous solution to the case of a line of single point thickness. The border concept would produce a result unacceptable to a subsequent edge-following routine, as the two boundaries of the line would be coincident.

A disadvantage of the edge as defined by Rosenfeld is that the resulting array of points has twice the point density of the original array, with no one-to-one relationship between the two arrays. A new array of edge points is therefore defined,

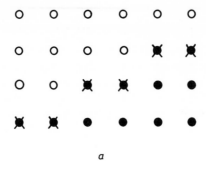

a

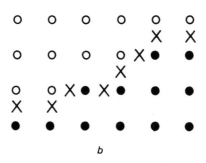

b

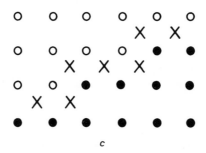

c

Fig. 3.   *Some possible outline definitions*

having the spacing and density of the original, but with each point shifted in both X and Y directions by one half-point spacing. Each point in the new array lies at the centre of a square formed by four points in the original array, and its state (black or white — edge or not edge) is derived from the states of these four points. A similar solution has been discussed by Saraga and Wavish.[12]

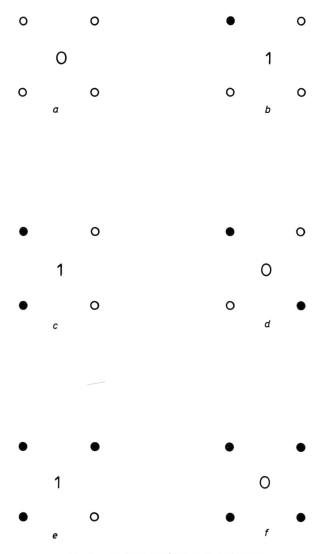

Fig. 4. *Definition of the outline elements*

The six basic configurations of the sixteen possible sets of four binary picture points are shown in Figure 4, together with the intuitively determined values of the associated points in the edge array. Configuration '*d*' is chosen not to yield an edge point in order to ensure the unambiguous treatment of single point thickness lines.

4. Edge Following – Procedure EDGETRACE

4.1 Parameters
EDGETRACE has been written to provide the basic function of determining whether a point located in the differentiated FRAME forms part of the outline of an object. The coordinates of the initial point are loaded into the global integers START X, START Y. EDGETRACE is then called as a conditional procedure, that is in a statement of the form

IF EDGETRACE *THEN. . . . . ELSE*

The 'condition true' return from the procedure occurs if a closed outline greater than a preset minimum length is successfully traced.

EDGETRACE provides a number of parameters concerning the outline, together with a description of the outline in the form of a list of vectors, stored as global variables, (Table 1). These parameters are illustrated in Figure 5.

Table 1   *Parameters Returned by EDGETRACE*

VARIABLE	VALUE
PERIMETER	The length of the traced outline (a positive integer)
AREA	The enclosed area (a positive integer)
XMAX XMIN YMAX YMIN	The maximum and minimum X and Y coordinates reached by the outline (negative integers)
XCENTROID YCENTROID	The X and Y coordinates of the centre of the enclosed area (negative integers)
CHAINCOUNT	The number of elemental vectors making up the traced outline; also equal to the number of outline points traced (a negative integer)
CHAIN	An ordered array of the directions of the vectors making up the outline. The array has CHAINCOUNT elements.

The generation of the values XMAX, XMIN, YMAX, YMIN has been made optional. If they are required, the flag word EDGEFLAG is set (that is, made non-zero). Similarly the recording of CHAINCOUNT and the array CHAIN may be allowed or suppressed using the word CHAINFLAG.

4.2 Chain notation and the basic algorithm
For the purposes of this study, a point in the image matrix is defined as being

connected to another point if it occupies one of the eight immediately adjacent locations in the matrix. The path between any pair of connected points may be denoted by one of eight elemental vectors (Figure 6) whose directions have been labelled -8 to -1 (negative to facilitate indexing in PL-516).

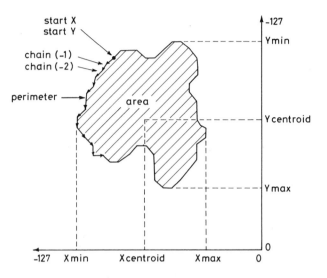

Fig. 5. *Parameters returned by procedure edgetrace*

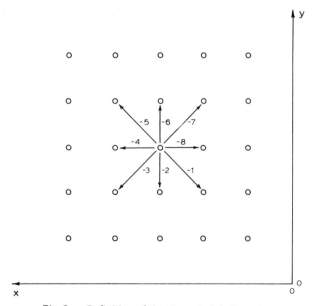

Fig. 6. *Definition of the elemented chain vectors*

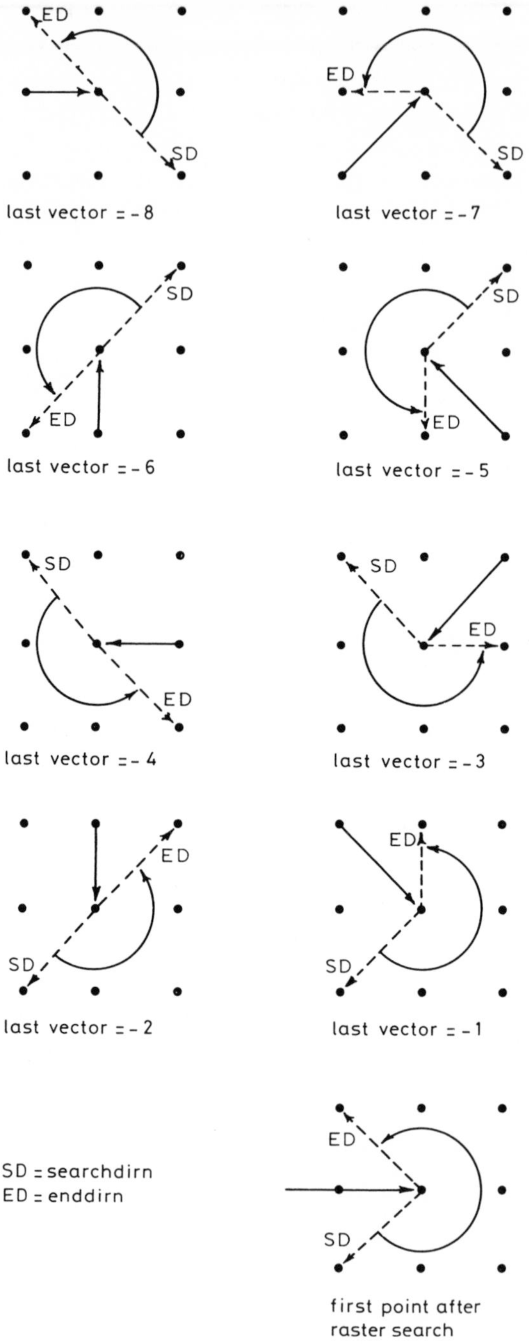
ED
SD
last vector = -8
ED
SD
last vector = -7
SD
ED
last vector = -6
SD
ED
last vector = -5
SD
ED
last vector = -4
SD
ED
last vector = -3
ED
SD
last vector = -2
ED
SD
last vector = -1
SD = searchdirn
ED = enddirn
ED
SD
first point after
raster search

Fig. 7. *New search and end directions*

An edge-following algorithm causes each point found in the outline to be reset to '0' before the search moves on to the next point. An outline is thus eliminated from the stored matrix FRAME as it is traced. The base for the search progresses around an outline in an anticlockwise direction, provided that the outline is first located from outside its enclosed area by the raster-type search. The outline can only be located from inside if it is intersected by the edge of the image matrix. In this case, the procedure EDGETRACE will execute 'condition false' return.

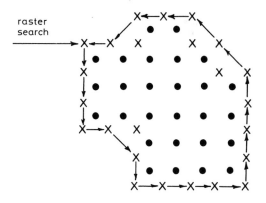

chain element number	chain element value
-1	-2
-2	-2
-3	-2
-4	-8
-5	-1
-6	-2
-7	-8
-8	-8
-9	-8
-10	-8
-11	-6
-12	-6
-13	-6
-14	-6
-15	-5
-16	-5
-17	-4
-18	-4
-19	-3
-20	-4

Fig. 8.   *The chain description of a simple outline*

From each base position not all the eight adjacent points need be tested, as some will have been tested from the previous base position. Before starting a local search, the first direction to be tested (SEARCHDIRN) is preset to value depending on the previous vector. The final direction to be tested before the search is abandoned (ENDDIRN) is also preset.

The present directions are shown in Figure 7, together with the values for the case of the first search after entering the routine, when the previous search would have been in a raster pattern. It can be seen that up to six points have to be tested in each search. However, the number of tests will normally be smaller. In the case of a straight section of outline, only one or two points will be tested from each base position before the next outline point is found.

A complete description of the outline is formed by listing the directions of the vectors linking one point to the next, in the order that they are determined (the array 'CHAIN'). The result of applying the edge-following procedure to a simple outline is shown in Figure 8, together with the derived 'chain' description.

As each element in an outline is located calculations leading to the determination of perimeter, area and first moments of area are updated.

4.3 Calculation of perimeter

The perimeter is measured as the sum of the magnitudes of the constituent elemental vectors. The vectors have magnitudes of 1 unit (even-numbered directions) or $\sqrt{2}$ units (odd-numbered directions).

The image matrix is assumed to be of unit spacing (Figure 6). As each point in the outline is located, one of the two integers EVENPERIM or ODDPERIM is incremented, depending on the last vector direction. At the end of the procedure the calculation

$$\text{PERIMETER} = \text{EVENPERIM} + \text{ODDPERIM}. \sqrt{2}$$

is performed, the result being rounded to the nearest whole number.

4.4 Calculation of area

The area enclosed by an outline is measured by summing the areas between each elemental vector and a convenient line (chosen to be the line $Y = 0$). The sign convention adopted results in vectors having a decreasing x component (directions -5, -4, -3) contributing positive areas, and vectors having an increasing x component (directions -7, -8, -1) contributing negative areas (Figure 9 – note that Y is a negative number). The net area enclosed by an outline traced in an anti-clockwise direction is then positive.

A simple example is shown in Figure 10. To avoid unnecessary manipulation of the factor $\frac{1}{2}$, the procedure EDGETRACE accumulates a total equal to twice the enclosed area (AREA2), and divides this by two in the final calculations.

4.5 Calculation of first moments of area and centroids

A similar method is used to calculate the first moments of the enclosed area about the axes $x = 0$ and $y = 0$. Figure 11 shows the eight elemental vectors, and the first moments of area of the areas between each vector and the x axis (ΔM_x). The sum of all the ΔM_x components for a closed outline will yield a negative total which, when divided by the enclosed area, will give a correctly negative value for the y coordinate of the centre of area.

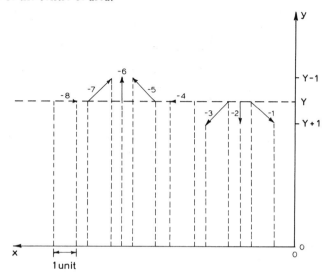

vector direction	Δ area	2Δ area
-8	$Y.1$	$2Y$
-7	$(Y-\frac{1}{2}).1$	$2Y-1$
-6	0	0
-5	$(Y-\frac{1}{2})(-1)$	$-2Y+1$
-4	$Y.(-1)$	$-2Y$
-3	$(Y+\frac{1}{2})(-1)$	$-2Y-1$
-2	0	0
-1	$(Y+\frac{1}{2}).1$	$2Y+1$

Fig. 9. *Element contribution for area calculation*

The expression for ΔM_x relating to odd-numbered vectors contains a constant term, $+\frac{1}{6}$ or $-\frac{1}{6}$. In a closed outline containing a large number of vector elements, each vector may be expected in approximately equal numbers, leading to a cancelling of the constant terms. The constant term will also be small compared with the Y^2 term and the enclosed area, particularly in the case of the close-up view of an object which then largely fills the frame. The calculation of moment of area is therefore simplified by neglecting the constant term. As for the area calculation, the factor $\frac{1}{2}$ is avoided by summing $2\Delta M_x$. The total is accumulated in the double length integer XMOMENT by the procedure SUMXMOMENT.

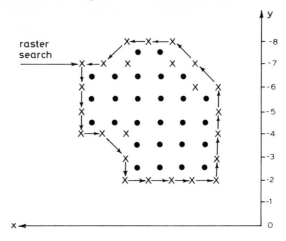

chain element number	chain element value	2Δ area	$\Sigma 2\Delta$ area
-1	-2	0	0
-2	-2	0	0
-3	-2	0	0
-4	-8	-8	-8
-5	-1	-7	-15
-6	-2	0	-15
-7	-8	-4	-19
-8	-8	-4	-23
-10	-8	-4	-31
-11	-6	0	-31
-12	-6	0	-31
-13	-6	0	-31
-14	-6	0	-31
-15	-5	$+13$	-18
-16	-5	$+15$	-3
-17	-4	$+16$	$+13$
-18	-4	$+16$	$+29$
-19	-3	$+15$	$+44$
-20	-4	$+14$	$+58$

Fig. 10. *The area calculation for a simple outline*

At the end of the procedure EDGETRACE, the total XMOMENT is divided by AREA2 and the result rounded by giving the centre of area Y coordinate YCENTROID.

Summing moments of area about the y axis gives the value XCENTROID in the same way. The contributions of each vector to this total are shown in Figure 12.

A single noise point, or a small number of adjacent points representing for example a piece of swarf, will give outlines of a few vectors only. A suitable value for MINCHAIN is found to be -10, that is outlines must have more than ten component vectors to be worthy of further consideration.

4.6 The image matrix after EDGETRACE

The edge-following algorithm employed resets each point it visits in the differentiated frame. All the points on an outline are not necessarily visited and consequently reset by the algorithm. Points set to '1' that are then left in the

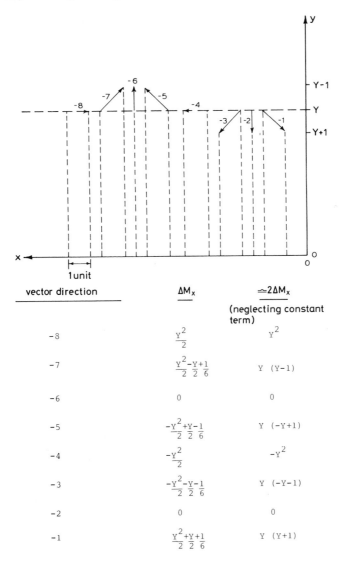

vector direction	ΔM_x	$\simeq 2\Delta M_x$
		(neglecting constant term)
-8	$\dfrac{Y^2}{2}$	Y^2
-7	$\dfrac{Y^2}{2}-\dfrac{Y}{2}+\dfrac{1}{6}$	$Y\,(Y-1)$
-6	0	0
-5	$-\dfrac{Y^2}{2}+\dfrac{Y}{2}-\dfrac{1}{6}$	$Y\,(-Y+1)$
-4	$-\dfrac{Y^2}{2}$	$-Y^2$
-3	$-\dfrac{Y^2}{2}-\dfrac{Y}{2}-\dfrac{1}{6}$	$Y\,(-Y-1)$
-2	0	0
-1	$\dfrac{Y^2}{2}+\dfrac{Y}{2}+\dfrac{1}{6}$	$Y\,(Y+1)$

Fig. 11. *Element contributions for X moment calculation*

frame will not normally form another complete outline, but they will constitute noise points during subsequent operations and care must be taken to minimise their effect. Figure 13 shows a plot of the image matrix FRAME of Figure 1 and 2 after EDGETRACE has traced both outlines, leaving the residual noise points.

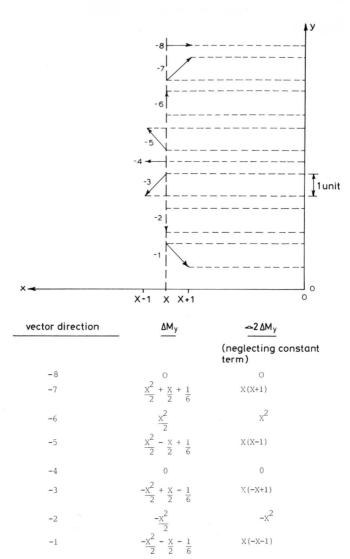

vector direction	ΔM_y	$\sim 2\,\Delta M_y$ (neglecting constant term)
-8	0	0
-7	$\dfrac{x^2}{2} + \dfrac{X}{2} + \dfrac{1}{6}$	$X(X+1)$
-6	$\dfrac{x^2}{2}$	x^2
-5	$\dfrac{x^2}{2} - \dfrac{X}{2} + \dfrac{1}{6}$	$X(X-1)$
-4	0	0
-3	$-\dfrac{x^2}{2} + \dfrac{X}{2} - \dfrac{1}{6}$	$X(-X+1)$
-2	$-\dfrac{x^2}{2}$	$-x^2$
-1	$-\dfrac{x^2}{2} - \dfrac{X}{2} - \dfrac{1}{6}$	$X(-X-1)$

Fig. 12.   *Element contributions for Y moment calculation*

4.7 Hole identification – procedure INSIDE

INSIDE determines whether a point located in the image FRAME lies inside or outside an established outline.

It is assumed that the outline generated by EDGETRACE is stored in the array CHAIN with the number of elements in CHAINCOUNT. The starting point of the chain is defined by the integers PARTXSTART, PARTYSTART.

The coordinates of the point to be tested are loaded into STARTX, STARTY and INSIDE is called as a conditional procedure. The 'condition true' return occurs if the test point is inside the stored outline. The stored values are not changed.

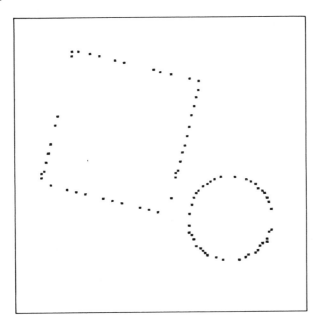

Fig. 13. *The frame of Figure 2 after both outlines have been traced*

The algorithm employed is based on the fact that in a plane, if a point P outside a closed curve S is connected by a straight line to another point Q, then Q is also outside S if and only if the line PQ intersects S an even number of times. If PQ intersects S an odd number of times, then Q is inside S (Figure 14). A proof is given by Courant and Robbins.[13]

In the procedure INSIDE a notional straight line is drawn from the point to be tested to the edge of the FRAME parallel to the x axis, that is the line $Y =$ STARTY, $X >$ STARTX (Figure 15). A pair of local integers (XCOORD, YCOORD) are initially set equal to the coordinates of the start of the stored outline. The elemental vectors defining the outline are read sequentially from the array CHAIN and used to modify XCOORD, YCOORD (procedure UNCHAIN) so that, XCOORD, YCOORD step through the coordinates of each point traced in the original outline.

At each step the original outlines are tested to see if the point lies on the line $y =$ STARTY, $x >$ STARTX. The number of times this line is crossed is counted in the integer CROSSCOUNT. Multiple crossings must not be recorded when the outline lies on the test line for a number of consecutive points (Figure 16). When the test line is met, a flag word is set, and the vector direction on meeting the line is noted (STRIKEDIRN). For each subsequent outline point lying on the test line, no action is taken. When the test line is left, the current, or leaving direction is compared with STRIKEDIRN to determine whether the outline has crossed the test line

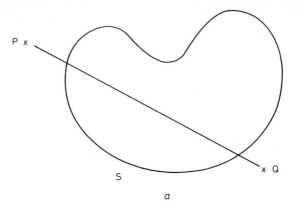

(a) Q outside S — PQ intersects S an even number of times.

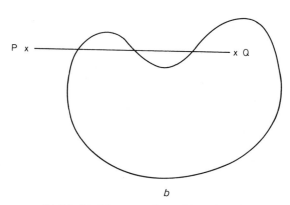

(b) Q inside S intersects S an odd number of times.

Fig. 14. *The basis of the inside algorithm*

(CROSSCOUNT is incremented) or only run tangentially to it (CROSSCOUNT is not incremented).

When each point represented by the stored outline has been regenerated and

tested, CROSSCOUNT is examined and the procedure exists accordingly. The potentially ambiguous case of the tested point lying on the stored outline cannot occur, as the action of tracing the outline resets all the points visited.

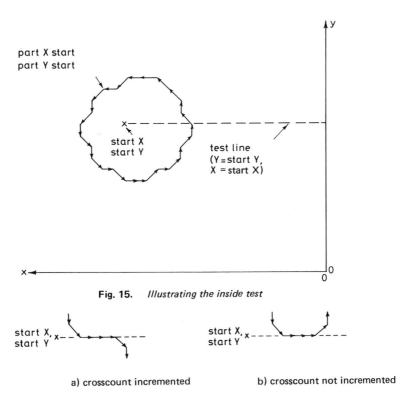

Fig. 15. *Illustrating the inside test*

a) crosscount incremented   b) crosscount not incremented

Fig. 16. *The tangent test for procedure inside*

It is verified that the object is correct, then its position and orientation must be determined to enable the machine to manipulate it as required by the assembly process.

5. The problem of orientation

The basic image processing procedures provide an analysis of outlines in the image (including both objects and holes in objects) in terms of area, perimeter, position of centre of area, enclosing rectangle, and chain vector description. These parameters will provide a useful first test, but fuller use of both hole and outline information is required to verify identity and determine orientation.

5.1 The holes model
The hole pattern in an object can be modelled in terms of the distances of the

centres of area of the holes from the centre of area of the object, and their relative angular positions. In addition, the areas and perimeters of the holes can readily be measured, yielding the data illustrated in Figure 17.

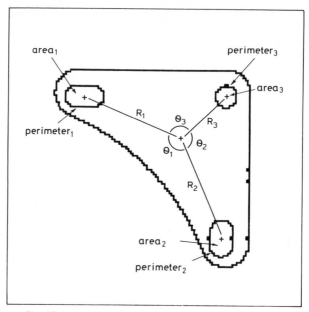

Fig. 17.   *Illustrating the parameters for the holes model*

This model is independent of the orientation of the parent objects, and can be directly compared with a stored reference model. When the individual holes in the scanned object have been related to the holes in the reference model, the absolute angular positions of the holes in the scanned object can be used to determine the orientation of the object in the input image.

5.2 Outline feature extraction

A model of the outline shape is immediately available in the form of the chain vector list. Comparisons of reference models and input data in this form as used for example by Freeman[7] are however complex and time consuming. A major difficulty is that the starting point of the chain on the outline of an object is arbitrary, depending on the orientation of the object and its position in the raster scan. The chains to be compared have therefore to be rotated relative to one another whilst searching for a match, and the chain description is not well suited to rotation operations other than in increments of $45°$. The correlation is further complicated by the fact that the reference and input are unlikely to have the same number of constituent elements.

A number of techniques have been developed for smoothing chain lists and normalising their lengths to facilitate comparison. Zahn[14] and Montanari[15] described methods of deriving minimal length straight line approximations to reduce the amount of data to be processed. Such techniques are unnecessarily complex in the context of an assembly machine, however. The requirement is to verify the tentative identification of the component, and to determine its orientation rather than to perform 100% inspection by detailed comparison of input and reference outlines.

The need for a model which allows simple, fast matching independent relative orientation suggest that an approach based on feature extraction from the outline may be more appropriate. Many character recognition methods, for example Hosking[16] exploit the fact that the line structure of the characters produces a set of common features such as line starts, joins, reversals and loops. Some corresponding features that could be defined for solid objects are sudden changes in outline direction (corners) and points of maximum and minimum distance of the outline from the centre of area. However, the unconstrained range of shapes that has to be accommodated by the assembly machine limits the usefulness of a technique based on a preprogrammed list of standard features.

A simple, interactive method of feature definition is required, whereby, during the learning phase, points on the outline of an object that are significant either as recognition or orientation criteria, can be specified by a human operator.

5.3 The circles model

The angular positions of the maximum and minimum length radius vectors are attractive recognition and orientation criteria, being readily derived from the data generated by EDGETRACE. However, if the length of the radius vector changes only slowly around the maximum and minimum values as the outline is followed, as for example in Figure 18 (a), the positions determined for R_{MAX} and R_{MIN} will be unreliable. This suggests that a more fruitful approach may be to specify a value of radius R, during the learning phase, for which clearly defined vector positions may be determined. In effect a circle of given radius is superimposed on the object, centred on the centre of area, and the intersection of this circle with the outline are defined as feature points. In Figure 18 (b) the feature points are defined by the parameters R, θ_1', θ_2, θ_3, θ_4. The angles may be differenced to produce an orientation invariant set of parameters for identification, and the absolute values used to measure orientation.

The insertion points are determined from the data produced by EDGETRACE by computing the distance of each point on the outline, represented by the chain list, from the centre of area. The points where this distance equals the specified radius are noted, and their angular positions from the centre of area calculated. The intersection points are thus found as an ordered list, the point with which the list starts being determined by the position on the outline at which EDGETRACE was called. The comparison of the input and reference lists of intersection points may thus require the rotation of one list relative to the other to find a match. Even before the lists are compared, the number of intersections found provides a useful recognition feature.

Suitable criteria for choosing the radius of the circle are:

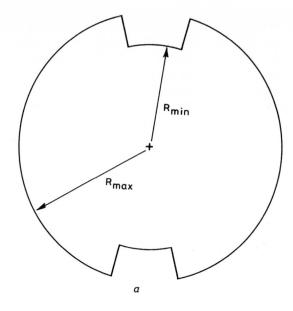

Fig. 18 (a)   *Illustrating maximum and minimum radius vectors*

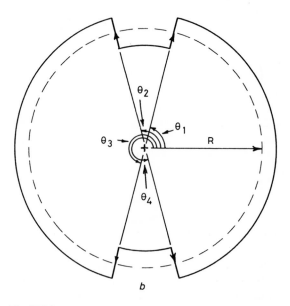

Fig. 18 (b)   *Illustrating the parameters for the circles model*

(i) the intersection points should uniquely define the orientation of the object

(ii) the number of intersection points should not be too large (say $\leqslant 8$)

(iii) for accuracy in measuring orientation, the radius should be as large as possible

(iv) the angular positions of the intersection points should not be significantly affected by small changes in radius

(v) intersection points should not vanish or new points appear for small changes in radius

The last two criteria are necessary to minimise the effects of small changes in the input image data.

In order to define uniquely the orientation of a component, or to differentiate it from another similar part, it may be necessary to specify more than one circle. The interesection points found when the outline is traced round will then form an ordered list of radius-angle pairs. In the example shown in Figure 19 two radii are used. The resulting radius-difference angle lists are shown for an object in two different orientations, the list entries being in the order they are found. To compare the two images, the two lists are first checked to ensure that they have the same number of entries. They are then compared for a match in both radius and relative angle entries. If no match is found, one list is rotated and the comparison is repeated. However, since both lists were generated using the same radius data, the radius values may be replaced by integer radius numbers.

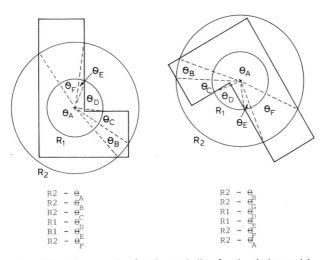

Fig. 19.   *An example of radius-angle lists for the circles model*

The measured angle values need only be compared if an exact match is found between the two lists of radius numbers, thus simplifying the comparison process. When the correlation between the two lists is established, the difference between the absolute angular positions of each pair of corresponding intersection points can be calculated. The mean of these six values is a reliable measure of the orientation of one object relative to the other. A further property of the circles model is the ability to distinguish the 'wrong way up' situation. Figure 20 shows that with the arrays rotated to give corresponding radius numbers, the angle difference values are in reverse order for the 'wrong way up' component.

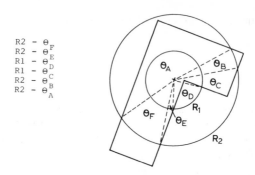

$$
\begin{array}{l}
R2 - \theta_F \\
R2 - \theta_E \\
R1 - \theta_D \\
R1 - \theta_C \\
R2 - \theta_B \\
R2 - \theta_A
\end{array}
$$

Fig. 20.   *The component of Figure 18 shown the wrong way up*

The circles model thus provides a powerful method of specifying features for recognition or orientation purposes on the outline of an arbitrarily shaped object. The amount of storage required for the reference data on one part is only two words per intersection point, and the corresponding data for the scanned part can be derived readily from the results of the EDGETRACE procedure. In conjunction with the holes model, it has therefore been used to provide the recognition and orientation capability of the assembly machine at the lower level.

Concluding remarks by lecturer

Readers must regard these notes as introduction only. A great deal has been published in the last decade in this area — particularly at international conferences. Some of this recent work is included elsewhere in the school notes.

Readers seeking further information on the visually interactive robot SIRCH referred to in these notes are advised to retrieve existing published material.[17-20]

References

1   ULLMAN, J. R; Pattern recognition techniques; Butterworths, London 1972
2   ROSENFELD, A; Picture processing by computer; Academic Press; New York 1969
3   BARROW, H. G. and POPPLESTONE, R. J; Relational descriptions in picture processing; Machine Intelligence 6; Edinburgh University Press, 1971
4   FREEMAN, H; On the encoding of arbitrary geometric configurations; IRE Trans. on Electronic Computers, June 1961

5   FREEMAN, H; Techniques for the digital computer analysis of chain-encoded arbitrary plane curves; *Proc. Nat. Electronics Conf.,* Vol 17, Chicago 1961

6   FREEMAN, H; On the digital computer classification of geometric line patterns; *Proc. Nat. Electronics Conf.,* Vol 18, Chicago 1962

7   FREEMAN, H; Apictorial jigsaw puzzles: the computer solution of a problem in pattern recognition; *IEEE Trans.* EC-13, April 1964

8   FREEMAN, H; A review of relevant problems in the processing of line-drawing data; Automatic Interpretation and Classification of Images, Academic Press, New York, 1969

9   DINEEN, G. P; Programming pattern recognition; *Proc. Western Joint Computer Conf;* Los Angeles 1956

10   UNGER, S. H; Pattern detection and recognition; *Proc. I.R.E,* October 1959, pp 1737–1752

11   ROSENFELD, A; Connectivity in digital pictures; *J. Assoc. Computing Machinery;* Vol 17, No 1, January 1970

12   SARAGA, P and WAVISH, P. R; Edge coding operators for pattern recognition; *Electronics Letters,* Vol 7 No. 25, 16 December 1971

13   COURANT, R and ROBBINS, H; What is mathematics?; Oxford University Press; London 1941, pp 267–269

14   ZAHN, C. T; A formal description for two-dimensional patterns; Proc. Joint International Cont. Artificial Intelligence, Washington 1969

15   MONTANARI, U; A note on minimal length polygonal approximation to a digitized contour; *Comm. ACM* Vol 13, No. 1, January 1970

16   HOSKING, K. H. and THOMPSON, J; A feature detection method for optical character recognition, IEE/NPL Conf. on Pattern Recognition; London, July 1968

17   Visual feedback applied to programmable assembly machines; Proc 2nd Int. Symp. on Industrial Robots; ITTRI; Chicago May 1972

18   HEGINBOTHAM, W.B., GATEHOUSE, D. W., PUGH, A., KITCHIN, P. W. and PAGE, C. J; Proc. 1st Conf. on Industrial Robot Technology; University of Bottingham, March 1973

19   PAGE, C. J; Visual and tactile feedback for the automatic manipulation of engineering parts; Ph.D Thesis, University of Nottingham, 1974

20   PAGE, C. J. and PUGH, A; Visually interactive gripping of engineering parts from random orientation; *Digital Systems for Industrial Automation,* Vol 1, No. 1, pp 11–44, 1981

Chapter 8

Three-dimensional imaging

C. J. Page

1. Introduction

The work described here investigates the use of scanning laser range-finders as the basis of a three-dimensional robot vision system. The types of components under consideration are those of complex shape and with features such as small blind holes which are difficult to detect reliably using simple vision systems which rely on a binary or silhouette image.

The overall operation of the system is as follows. Parts are presented to a computer-controlled sensing head on a feed track from a conventional parts feeder. The component receptacle of the sensing head is a flat, transparent plate on which the part comes to rest in a random position and orientation and in any of its stable attitudes. The component is imaged simultaneously from each of the six sides required to perform a complete inspection of it from all directions. A set of six identical sensors are used, each of which produces an image of the part from its own particular viewpoint. The data from the sensors is then analysed and correlated by the control microprocessor to yield a representation of the component's geometric shape. Further analysis recognises the part and computes its orientation by comparison with a preprogrammed version showing it in its preferred attitude.

Most of the current robot vision systems employ a sensing device (usually a television camera of some form) which generates a two-dimensional image of the scene or objects before it. If depth information is required, ancillary sensors must be used. However, other workers have reported systems in which three-dimensional imaging is performed directly. Triangulation methods are most commonly used,[1] although researchers at SRI have reported work on optical radar techniques applied to the imaging and analysis of office scenes.[2,3] Descriptions of research on the use of tactile methods for sensing three-dimensional engineering components have also been published.[4]

Each of the sensors in this application generates a three-dimensional image of the workpiece from its own particular viewpoint by measuring the range to the part over a square matrix of sampling or picture points. Complete, unambiguous data on the workpiece is obtained, and this data is unaffected by variations in the angle and level of illumination. The method used for rangefinding is similar to that used by the SRI workers and measures range by time-of-flight measurements of a beam of laser light reflected back to its source by the component. This method is being employed in preference to triangulation because the latter can produce blind spots at the abrupt discontinuities in profile which occur in many engineering components.

2. The imaging system

2.1 Introduction
The overall arrangement of the inspection station is shown in Fig. 1.

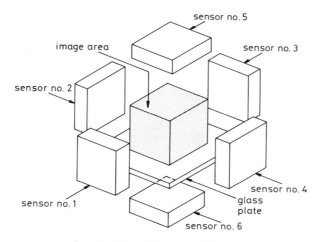

Fig. 1   *Schematic of inspection station*

Six identical sensors view the component from each side. Each sensor scans a square field of view with a television-like raster scan of 64 lines per frame. The range to the object is sampled 64 times along each line to produce a 64 × 64 square matrix of range values for each of the six view points. An important feature is that the scanning beam is always normal to the image place. This produces a constant size image free from perspective distortion. The part under scrutiny rests on a flat, anti-reflective coated, optical glass plate. Reflection and refraction of the image seen by sensor 6 through the glass plate is minimised by the normal scanning arrangement. The six sensors are arranged so that their fields of view intersect to generate an imaged volume in the form of a cube. The sensor dimensions are such that this cube has sides of length 100 mm. Range resolution is chosen to be 1.5625 mm to divide the image volume into 64 × 64 × 64 elemental cubes. Range over the depth of the image volume is stored as a 6-bit binary number. Using a byte of memory for each elemental picture cube or pixel means that 4 kilobytes of memory are required for each sensor, or 24 kilobytes for the set of six.

A scan time of 2 seconds has been chosen, somewhat arbitrarily, for the proto-type system. This allows approximately 500 microseconds between the start or successive range samples. It is anticipated that further development will allow a scanning time of 0.5 seconds or less to be achieved without significant performance degradation.

2.2 Range detection

The method used for measuring range to the object under scrutiny relies on detecting the light reflected from a small, intense spot of light projected onto the surface of the component. The light detector is positioned so that it measures scattered light in a direction coaxial with the transmitted beam, thereby eliminating blind spots which can be encountered using triangulation methods. Because of the coaxial arrangement of transmitter and detector, the latter relies in general on picking up radiation from diffuse rather than specular, or mirror-like reflection at the surface.

Simple experiments with a low power laser and a silicon photodiode-amplifier combination shown the method to be viable for common engineering materials. As is to be expected, the detector output peaks at near normal angles of incidence and decreases rapidly as the angle increases and diffuse reflection becomes the dominant mechanism. However, it has been found possible with even the simple experimental arrangement used to detect reflected light at angles of incidence in excess of 80° to the normal.

The surface texture of the component is an important parameter as the method is obviously not practical for surfaces with truly mirror finishes. In these cases, range is measured to the object 'seen' in the mirror. Nevertheless, even for metals with 'good' finishes produced by processes such as polishing, grinding, and finish milling and turning and with CLA values of the order of 0.1 microns, the exper-iments show a detectable response at angles of incidence in excess of 80°. The main problem is the large dynamic range-up to 80 dB even before the effect of range variations is taken into account.

Material composition affects the amount of reflected light too, as non-metals such as plastic and wood are not as reflective as metals. For these materials, the dynamic range of the detected signal is not so great as for metals of similar surface roughness, and mean signal levels are somewhat lower. Materials such as glass and transparent and translucent plastic provide the other main limitations to this imaging method as scattered light from both front and rear surfaces is detected, yielding an average range value.

For this type of application, the choice of lightsource and detector is particularly important. Ideally, both should be small, robust, efficient, low-power devices. For the light source for each sensor a continuous wave (CW) GaAlAs (gallium aluminium arsenide) solid-state, double heterostructure laser diode of 15 milliwatts maximum optical output power has been specified. The wavelength of the light output is 800 nano-metres, which is in the infrared. The detector for each sensor is a silicon avalanche photodiode. This is a solid-state device which incorporates internal gain (around 200 typically).

Range is measured by modulating the continuous wave output from the laser diode with a sinusoidal waveform and measuring the phase shift over the path length of the light beam, which is twice the range. This technique has a number of advantages over the alternative pulse-elapsed-time method often used for long-range instruments. Firstly the inherent turn-on or lasing delay of one or two nanoseconds is reduced to zero by dc biasing the laser diode and superimposing a sinusoidal modulating waveform. Secondly, in the pulsed system the fast edges in the reflected light can be masked by the inevitable pick-up and noise. The effects of noise can be reduced in the CW system by passing the detector output through a narrowband filter tuned to the modulation frequency. Thirdly, the detector output in the CW system can be mixed to generate an intermediate frequency (IF) signal with the same relative phase shift as the original, but at a much lower and more tractable frequency.

The resolution required from the range detection system is approximately 1.5 mm. At a modulating frequency of 50 MHz, for example, this gives an incremental phase shift of $0.2°$. While it is possible to measure much smaller values of phase shift, in view of the low signal levels and also the wide dynamic range of the signal it has been considered advisable to use a modulating frequency in the tens of megahertz range.

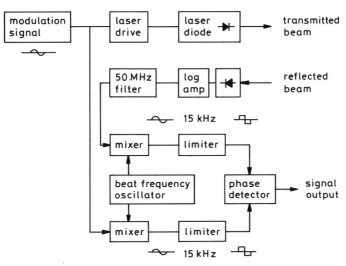

Fig. 2   *Block diagram of range detection system*

A block diagram of the electronic control system of the rangefinder is shown in Fig. 2. The 50 MHz sinusoidal modulating signal is derived from a crystal controlled oscillator and modulates the laser diode bias and drive circuit directly. The laser diode transmits a beam of infrared light which is modulated at 50 MHz with as high a modulation factor as possible to improve the signal-to-noise ratio at the detector.

Reflected infrared light is picked up and amplified internally by the silicon avalanche photodiode detector. A logarithmic amplifier is used in the following amplification stage to compress the wide dynamic range of the incoming light levels. The output from the amplifier is then filtered by a narrowband filter to reduce unwanted noise levels. A beat frequency oscillator is used to generate an intermediate frequency output of about 15 kHz from both the initial, reference modulation signal and the output from the filter. After passing through a limiting stage, the phase difference between the two signals is detected and converted to digital form ready for interfacing to the control processor. Offsets due to differential phase shifts in the electronic system and the varying optical path length through the scanning system are dealt with in a preprocessing operation in which suitable values stored in lookup table form are subtracted from the digital output of the phase detector.

2.3 Scanning system

The beam scanning system uses a single transmitter and detector per sensor with a mechanical/optical arrangement performing the scanning operation. This method has been adopted in preference to arrays of transmitters and receivers because of the high cost of suitable devices. A schematic diagram in plan view of one possible scanning system is shown in Fig. 3.

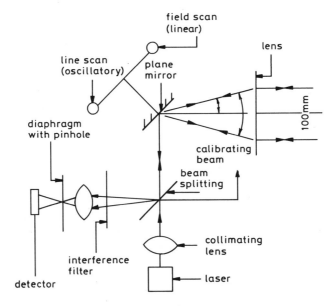

Fig. 3   *Schematic of scanning system*

A normal scanning beam is produced by reflecting the laser beam off a plane mirror located at the focal point of a large convex or aspheric lens corrected for spherical aberration. Rotating the mirror in the plane of the diagram scans the beam horizontally across the image plane. An alternative method which is also under consideration uses a large parabolic mirror instead of a convex lens, again with the plane mirror at its focal point.

That part of the reflected radiation which is coaxial with the transmitted beam returns along the same path, and a fraction of it is deflected by the beam splitter. It then passes through a narrowband interference filter with a passband of approximately two nanometres, whose purpose is to reject radiation from all sources other than that particular sensor. In practice, the most troublesome sources of radiation are the other sensors. This problem is overcome by using a laser diode of slightly different wavelength for each sensor, each with its own interference filter. The light output from the interference filter is then focused by a lens system onto a pinhole in a diaphragm. This limits the maximum off-axis deviation of the reflected light to less than 0.5 degrees (depending on the size of the pinhole). The light then impinges on the surface of the avalanche photodiode.

Line scanning is achieved by oscillating the plane mirror by means of a proprietary optical scanner on which it is mounted. This is a small, precision moving-iron galvanometer device with a built-in capacitive transducer which permits accurate control of both mirror angle and angular velocity. Vertical, or field scanning is performed by incrementing the whole assembly shown in Fig. 3 downwards by approximately 1.5 millimetres at the end of each line scan using a stepper motor drive operating in open loop mode under computer control. Only a small section of the lens or parabolic mirror is then required. This method eliminates the pincushion distortion of the raster that would be inevitable (unless the lens or mirror were designed to compensate for it) if the mirror were tilted to produce the field scan.

Accurate spacing of the range samples along each line is achieved by using that part of the transmitted laser beam which is unavoidably reflected by the beam splitter (the 'calibration beam' in Fig. 3). This beam is redirected by a suitable arrangement of prisms and/or plane mirrors back into the line of the main beam, but displaced vertically by a small amount. The calibration beam strikes the plane mirror and is rotated through the same angle as the main beam. It then passes through a reticle of 64 equally spaced dark lines, and through a lens which focuses it onto a silicon photodiode – amplifier combination. A series of sampling pulses are therefore generated from the photodiode amplifier which are used to synchronise range sampling along each line of the scan. Thicker lines at each edge of the reticle can be used to generate synchronisation pulses for the vertical, or field scan mechanism.

3. Image processing algorithms

The purpose of the image processing algorithms is to recognise the imaged component and to compute its orientation with respect to that of a preprogrammed reference part whose parameters are held in memory. Whilst in the ideal case there is only one object in the viewing station at a time, it is nevertheless desirable that the algorithms be capable of coping with a 'scene' consisting of more than one component.

The raw data from the imaging system consists of six blocks of sixty-four lines of sixty-four range samples per line, with each sample occupying one byte of memory. A certain amount of preprocessing will have been done to eliminate differential optical and electrical effects and to remove reflections from objects not within the image volume (for example, the sensor opposite).

The first major processing operation is to combine the individual blocks of sensor data into one data structure. The characteristics of the imaging system and also the exact edgewise registration of the six image planes one with another means that correlation is an inherent feature. The data structure chosen is a simple one to facilitate rapid processing; namely a three-dimensional matrix of bits in which every logic one in the matrix represents a surface point on the object and its position in the matrix the three-dimensional coordinates of that point.

To realise an industrially viable system, the image analysis software must be rapid in operation. Because of the comparatively large amount of data involved, the algorithms employed must be computationally simple; and in keeping with this philosophy, the first attempts at algorithm design have used an analogous approach to current two-dimensional binary imaging systems[5,6] where recognition is carried out by computing a set of numerical shape and size descriptors and comparing them with sets of reference data obtained during a previous 'teaching' operation. Suitable ones for the three-dimensional case are surface area, volume, maximum and minimum radii from the centre of volume to the surface, moments of inertia, and the number of through and blind holes.

The analogy is taken one step further by processing the three-dimensional data structure as a set of sixty-four 'slices', where each slice is a one volume element thick section through the data structure parallel to one of the coordinate axes. Each slice is processed essentially independently as if it were a binary image of a laminar object of unit thickness, with running totals being accumulated for the descriptors where appropriate as the component slices of the data structure are examined in sequence. Extensions of binary two-dimensional algorithms are used where applicable. One candidate under consideration is the run-length coding algorithm[6] where each frame is scanned line by line noting zero-to-one and one-to-zero transitions only. The data can be different in the three-dimensional case, however, as each slice can be a 'blob'; that is, a region of ones at a surface parallel to one of the coordinate axes, a closed curve for a slice in the object interior, or a combination of one or more of each type.

A particular problem which arises is that of missing points in the surface model of the component. It is caused by the sensors not being able to 'see' parts of certain surfaces depending on the orientation of the object. These include the sides of holes and reentrant features such as notches and rebates, and surfaces occluded by another component. This means that some of the contours shown by the interior slices of the component will be incomplete or have unclosed spurs and it is necessary to 'fill in' these missing points.

The infill operation is necessarily the first one performed on the data structure and operates by comparing adjacent slices. The success of this algorithm depends on the viewpoint, so for ease of processing the data structure is reordered in two additional ways to give a total of three models consisting respectively of slices in plan view, from the side and from the front. Each of these models is processed in sequence, any infilled surface point in one model being set to one in the other two.

After this operation, the set of shape and size descriptors can be calculated and compared with the corresponding reference values. The mean and standard deviation are available for each of these, obtained during a training operation in which the reference part is imaged many times in different orientations and positions.

Recognition by comparison of numerical parameters is not accurate enough for reliable recognition discrimination between parts which differ by only small features, so final recognition is combined with orientation computation. The technique under consideration is an extension of the 'polar signature' method used for binary two dimensional images,[5] in which the radii of all edge points from the centroid are calculated. One or more particular radii are chosen for the reference part and the sequence of these and their angles forms a signature whose phase angle is proportional to orientation. In the three-dimensional case, the radii are not limited to two dimensions and as the model is processed slice by slice, the order of the radii values depends on the orientation and the polar signature does not have a constant shape as in the two-dimensional case. For recognition purposes, one way round this problem is to produce a frequency distribution of the radius values obtained and match this with a reference. For orientation computation, a radius or radii are chosen which have only a few definitive surface points represented. The corresponding set of orientations must then be correlated with the reference set to yield the component orientation.

4. Concluding remarks

The rangefinder imaging system described in this paper possesses several character-istics of importance to both automatic inspection in the wider context and to robot vision. It provides unambiguous precisely registered three-dimensional scene infor-mation from a number of view points. The data is free from illumination effects such as shading and shadows, does not suffer from perspective distortion, and con-tains errors due to reflections only when very highly polished surfaces are involved. The rangefinder could form the basis of a powerful robot vision system, with the number of individual sensors being specified according to the application.

It is believed that trying to develop simple binary-type algorithms for three dimensional part recognition is a fruitful avenue of investigation. The techniques are inherently fast in operation and successful development should achieve image processing times of one or two seconds. The way that they are structured lends itself to parallel processing, and further development towards implementation by a multimicroprocessor system could produce an improvement of at least an order of magnitude on these estimates.

5. Acknowledgement

The author gratefully acknowledges the financial support provided for this project by the UK Science and Engineering Research Council.

6. References

1. ISHII, M., and NAGATA, T. Feature extraction of three-dimensional objects and visual processing in a hand-eye system using laser trackers. Pattern Recognition (GB), Vol. 8, 1979, pp. 229−237.
2. NITZAN, D., BRAIN, A. E. and DUDA, R. P. The measurement and use of registered reflectance and range data in scene analysis. Proc. IEEE, Vol. 65, 1977, pp. 206−220.
3. DUDA, R. O., NITZAN, D., and BARRETT, P. Use of range and reflectance data to find planar surface regions. IEEE Trans. Patt. Anal. and Mach. Intell., Vol. PAM-1, No. 3, 1979, pp. 259−271.
4. PAGE, C. J., PUGH, A. and HEGINBOTHAM, W. B. New techniques for tactile imaging. The Radio and Electronic Engineer, Vol. 46, No. 11, 1976, pp. 519−526.
5. HEGINBOTHAM, W. B., GATEHOUSE, D. W., PUGH, A., KITCHIN, P. W., and PAGE, C. J. The Nottingham SIRCH assembly robot. Proc. 1st Conf. Industrial Robot Technology, Nottingham, UK, 1973, pp. 129−143.
6. GLEASON, G. J., & Agin, G. J. A modular vision system for sensor-controlled manipulation and inspection. Proc. 9th Int. Symp. on Industrial Robots, Washington DC 1979 pp. 57−70.

Chapter 9

Tactile sensing

M. H. E. Larcombe

1. Natural systems

The sense of touch is at once our most direct sense of the physical world and the least researched sense. It is in fact not one sense at all but a combination of at least four and depends for its effectiveness on the use of sensor signals within a control loop.

Nature's sensing elements comprise: two forms of pressure sensor, one for transient load detection, one for continuous load measurement; a form of position transducer giving the 'proprioceptive' data; overload and fault sensors falling also in to the realm of pain sensors, reacting to internal structure damage and to external chemical stimuli. We should also consider the temperature gradient sensors as part of the tactile system.

These sensors are liberally distributed over the entire skin surface. With such multiplicity of sensors we might expect some superb pattern recognition abilities. In fact the system by itself is almost devoid of pattern processing. Over large areas the ability to discriminate between two points requires a spatial separation of two centimetres or more. In addition we find that many groups of sensors share common 'data channels' even when the groups are widely separated — leading to some bizarre signal confusions such as an injury to the foot giving a sensation in the mouth. Even in those areas well provided with sensors such as the finger tips and the lips pattern recognition abilities are very limited. This is instanced by the difficulty experienced by the blind in learning the relatively simple patterns of six raised dots used in Braille. The sensors are however extremely sensitive detecting movements of only a few microns, even a cobweb contact can stimulate the cells (and it is notoriously difficult to establish the position of the stimulus).

Fingerprints, or rather the whorls and ridges raised from the finger tips, allow the assessment of surface texture when relative movement occurs. The motion causes a mechanical interaction between the ridges and the surface to produce a vibratory signal. Unless a texture is coarse enough to stimulate the pressure overload sense or has a marked thermal effect surface texture is virtually undetectable without relative movement.

If the sensors themselves appear to have such a poor intrinsic pattern recognition function how is it that we can use feel to recognise objects? Let us first note that we use another word for our use of touch, 'feel'. Feeling involves the use of the sensors to guide the probing and grasping actions of the fingers and hand. The sensors act within a sensorimotor control loop to provide the so called 'haptic' sense, a word meaning quite simply holding. In simple terms the sensors allow the accurate placing of the hand or other parts of the body against the surface of objects. The shape interpretation is then conducted using the position feedback from the hand through the proprioceptive sensors. We parse the object by physically moving from 'hapteme' (to coin a word) to 'hapteme' under a grammer based on our own geometrical mental models of objects and their components. Simple haptemes are such things as edges and corners and we use very simple servo following methods to explore these.

The overall skin sensor system serves many purposes but its role is predominantly protective. Although the skin cannot by itself avoid damage, not being self actuating, it can initiate reflex mechanisms to ameliorate the effect of impact or abrasion. Even more importantly it works in conjunction with deep body sensors to prevent self inflicted injury due to mechanical overload.

Unfortunately little seems to be known about the intrinsic mechanisms of the biological sensors. Even were we to know how the essentially microbiochemical devices function it is unlikely that we could emulate the mechanism at the macroscopic level. We do know something of their behaviour. The sensitive sensors appear to respond as rate of change devices while the load sensors appear to produce neural outputs related to the intensity of load. As with many biosensors the outputs have a logarithmic characteristic and have the property of fatiguing or reducing their 'gain' on prolonged stimulation. As measuring instruments they have low accuracy with about five bits range.

2. Technological systems

In principle it should not be difficult to produce a range of touch sensors matching those of biological systems. However until recently there were no available sensors suitable for use with manipulator systems. Technology has only recently been faced with the need to produce them.

Essentially the transient detection could be affected using small piezo-electric elements. The generation of the load signal presents more problems. The measurement of load has in the past generally involved the measurement of strain of an appropriate mechanical structure, using either LVDT, or resistive strain gauges. All such measurement techniques have been used with some success on manipulator structures. The necessary bulk of such measurement devices generally precludes their use in quantity on most manipulative systems. In addition the signals from devices such as strain gauges are very weak leading to problems from noise.

Much of the technology for load measurement was developed for accurate load measurement with much consideration of linearity and repeatability. For the purpose of robotic manipulation the requirements may be relaxed considerably. A number of research workers, including the author, have accordingly experimented with pressure sensitive resistive devices. Yet other workers have investigated capacitive devices. Some work has been done using pressure sensitive semi-conductor devices although these have the drawback of mechanical fragility.

The pressure sensitive resistors are frequently based on graphite loaded neoprene rubbers. The most promising technique (Purbrick 1981) uses the variation of resistance at the interface between two orthogonal rubber cords. The lack of robustness of the rubber elements precludes their use in some applications. The rubber does have a finite and relatively short life under repeated cycling — the proportion of carbon that must be included to render the rubber conductive reduces the mechanical durability.

The carbon fibre elements developed by the author (Larcombe 1981) are both sensitive and robust, possessing wide dynamic range. These make use of the change of resistance of a felt form of carbon fibre — the same principle as the rubber cord sensor but multiplied some million fold. Given appropriate felts excellent results may be obtained.

Not all technological techniques mimic biological sensors. Some years ago the author gave a gripper structure an overall tactile sense using a method based on vibration. A gripper is a mechanical structure and as such has its own modes of vibration. The technique involved exciting the principle mode using a small electro-magnetic exciter at the appropriate frequency. The phase of voltage and current through the exciter was monitored. Any contact with the structure produced initially 'rattle' noise across the exciter but more significantly a substantial phase change as the mechanical structure changed its resonant frequency. By tracking the resonant frequency as it fell it was possible to extend this 'tactile' sense to the workpiece as well! With some mechanical structures the frequency of resonance is such that a significant proportion of the energy is transmitted out as sound waves. When this is the case it is found that the presence of objects within the near acoustic field may cause sufficient energy to be reflected back to produce a change in the effective resonant frequency of the system. Such effects extended over ranges of less than 1 cm typically but nevertheless give a form of proximity detection.

The use of piezo-electric elements was touched upon earlier. Experiments have been made using arrays of such elements. Piezo elements must be understood to use them effectively in this application. Under pressure a voltage is generated across the element. If measured by an inifinte impedance device this voltage is constant for constant pressure. If, as is physically inevitable (by leakage through the element itself), a current flows the voltage decays and will result in a reverse voltage pulse on release. For this reason such elements are often purposely loaded resistively to act as transient or rate of change devices. Used as a load measuring device due consideration must be given to the element time constants.

As detectors of initial contact such devices are clearly useful although it must be remembered that as with all sensitive tactile sensors they are naturally microphonic (or geophonic or 'structure-o-phonic'). They thus serve as an excellent detector of load slipage, especially if the gripping surfaces are rough, as is often the case. With a positively gripped load they can be used to detect, through structural sound transmission, the contact of the load with another object. The signal levels of these structurally transmitted signals are well defined above background noise levels, provided steps are taken to isolate or reduce the transmission of noise from the manipulator drive mechanisms. The conventional structurally strong anti-vibration mounts serve well for this purpose.

In addition to the above devices mechanical switches provide a form of tactile detection. They may however present some problems of control in use (q.v.).

3. Operational considerations

For most industrial manipulation we are not concerned so much with the use of tactile sensing for object recognition. If we create a tactile sensor for object recognition or for object attitude recognition we are creating the input for a pattern recognition system and the problem is very similar to that of optical input. Instead we are concerned with the use of the sensors to assist gripping, placing and tracing.

3.1 Gripping

The use of load and transient sensors on the gripping surfaces allows the provision of adaptive gripping. If an open gripper is placed about the load the gripper may be slowly closed until first contact is made. This will occur preferentially on one side or another (in essentially sampled control such as we have with computer control the 'first' contact is always defined). The manipulator can then be servo compensated to ensure that the load is not moved until the second gripper surface makes contact. When the gripper is in contact the gripper may be closed until the gripping force reaches the level assessed for the load. Alternatively a cycle of attempted lifts and slippage monitoring may be entered until slippage ceases — such grip stabilising cycles could be maintained throughout handling.

This adaptive gripping works independently of the work size and is thus of great use in teleoperator systems. It is of course wise to ensure that 'grip' is not verified if the gripper is in fact fully closed — the gripper is empty! If the gripping load measured stays constant or even starts to fall as the gripper closes the work is either viscous or is breaking — again vital information for teleoperator systems.

The author has had great success with such adaptive gripping systems, allowing a heavy duty gripper to lift loads varying from an egg to a 54 kg load (the author).

3.2 Placing

The detection of work contact with another object has already been mentioned. By using load sensors to measure the vector load on the gripper sensitive placing operations may be made.

The classic placing problem is the task of inserting a stud in a hole. It is usually possible to cover part of the hole (presuming no hole in one situation). Contact with the surround is detected as mentioned above but now the load sensors detect the attempted movement of the stud about the common chord of the circles about the top of the hole and the bottom of the stud. This rocking motion is in the direction of the centre of the hole by moving incrementally in this direction and making repeated small insertion attempts progressively better estimates may be made of the centre.

3.3 Tracing
If we require to follow a circuitous mechanically demarked track the use of a tactile probe may be used. This is of particular interest in automatic welding where a poorly specified (in numerical terms) weld line exists. A tactile probe capable of sending the line — either directly or through the mechanical load on a mechanical probe — may, by servo-control, lead a welding head along the line. It is necessary to take into account the offset between probe and welding head! It is also necessary for stability of such tracing to incorporate a rate of change term. A high stability tracing system requires the predicted slope term at the point being traced.

4. The use of switches

For smooth control a form of continuous (not necessarily linear) sensing is desirable. It may be inconvenient or undesirable to provide the necessary sensors. Mechanical or conductive rubber switches (metal particles in a rubber matrix) may be the only convenient sensor. We might suspect that such binary sensors will give at best bang-bang control, undesirable at best. This is not the case. With a computer controlled system it is possible to incorporate forms of mathematical (or rather algorithmic) model of the work. In this case switch inputs provide confirmatory signals of events which are expected to occur. A control loop which is open as far as environmental feedback is concerned over most of its range is momentarily closed by the switch allowing instantaneous correction of the cumulative error.

5. Summary

'Tactile Sensing' is an inadequate description of the use of mechanical sensing in handling. The sensors themselves do not provide the data for control. Sensing occurs when a sensorimotor feedback is used and it is the signals from these feedback loops which provide the control data. The methods described provide readily accessible technology without the problems posed by pattern recognition or image processing. Since direct mechanical sensing is involved it is relatively easy to program systems using tactile feedback as the signals are very positive and do not require much interpretation.

References

PURBRICK, J. A. 'A Force Transducer Employing Conductive Silicone Rubber', Proc. 1st. Int. Conf. on Robot Vision and Sensory Control April 1981, I. F. S. Ltd.

LARCOMBE, M. H. E., 'Carbon Fibre Tactile Sensors', (ibid).

Chapter 10

Computer simulation to aid robot selection

M. Dooner

Abstract

The use of computer simulation to aid robot selection and evaluate robot work-station designs, provides the engineer with an invaluable tool for planning robot installations.

The ability of robots can be examined in terms of reach, geometry and cycle time, and task programs developed and de-bugged prior to installation.

A system named GRASP – Graphical Robot Applications Simulation Package – being developed in the department, will be described and a number of case studies will be presented to illustrate the potential of the system.

It is proposed that the system will be extended to allow robot task schemes to be developed completely off-line and the task program specification transferred directly via an interface to the robot controller. This proposal will be outlined.

1. Introduction

This paper describes a technique which applies interactive computer graphics to simulate the operation of industrial robots in the workplace.

The GRASP system – a development of the successful SAMMIE CAD software – produces models of robots and workplaces and displays them on a graphics terminal.

The benefits provided by using computer graphics simulation, are:

Rapid robot assessment and workplace design
Off-line assembly and evaluation of task schemes prior to installation
Motion simulation and computer animation.

The SAMMIE CAD software, although principally an ergonomic design tool, provides the general features offered by a geometrical modelling system. The basic

geometric modeller is augmented by a number of facilities that allow an industrial robot to be evaluated and a robot/workplace layout to be designed. These
include:

Robot library — a data bank of kinematic models
Robot Operating System — permits the designer to program and execute task
sequences.
The operating system contains a number of back-up programs to assist robot
evaluation:
Coordinate Transformation routine — transforms world coordinates into
specific joint coordinates.
Trajectory Calculation routine — provides the user with optional modes of
motion between programmed points.
Event Processor — a structured mechanism for the processing of tasks and
sub-tasks in a time ordered way.

2. Geometric modeller

The SAMMIE computer aided design software provides the basis and general
structure for the kinematic modelling system required of GRASP. The software
consists of a hierarchical data structure which provides online facilities for building,
displaying and manipulating 3D geometric models. The geometry of a model
is described using data which is generally prepared off-line using statements and
is then translated by the software into a structured data base which may be
amended interactively. The model is defined by the size and shape of its components,
and a detailed specification of their interconnections. The model components
must be simple convex polyhedra, but these may be combined to form complex
solids.

Homogeneous transformations represent the spatial relationships between different parts of the model and are stored explicitly and may be changed under
program control. These transformations are stored in a hierarchy which reflects the
successive application of transformations along a linked structure. This data structure ensures that when the base joint of a robot is moved then the 'end effector'
(tool point, say) moves in the correct way.

The user interacts with the model via a number of menus, which include facilities
for model viewing, displaying and manipulation. A hidden line removal routine is
available but is rather costly computationally, and is not practicable during the
displaying of motion, however, it is used for static illustrations.

The manipulator linkage mechanism is described through the matrix formulation. Each link is interconnected through a $[4 * 4]$ homogeneous transformation

matrix, shown below, where the link orientation is expressed by the rotation term '*R*', and a column vector '*T*' gives the location of the base of the link.

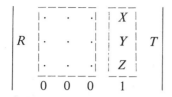

2.1 Robot models

A robot data file within the library contains sufficient information to allow the user to program and execute a task sequence, and to evaluate the robot's capabilities with regard to its kinematic performance. Geometrical capabilities refer to (a) the spatial coverage of the end effector, usually the tool point and (b) the tool's (or gripper) orientation capabilities.

A description of the mechanism's geometry includes coded matrices which identifies whether the joint is revolute or prismatic. Joint constraint data determines each robot's working envelope. The robot models can be displayed in either stick form with the 'flesh' removed, or a fully representative geometrical model — important for visual presentation.

Robots contained in the library are assembled from geometrical data usually obtained from manufacturers literature. The library contains most industrial robots found in commercial use.

3. Manipulator kinematics and control

The kinematic model describes the geometry and motion of the manipulator, without reference to forces. Dynamics are not considered here. Coordinated control allows the user to position and specify the rate of motion of the manipulator end point in the workplace.

Manipulators with up to six degrees of freedom are modelled by the simulation system. This is a minimum for total spatial command where, three joints position an object in a desired location, and three local joints provide object orientation. Naturally, extra degrees of freedom provide extra manoeuvrability, but presently the system is restricted to six d.o.f. manipulators, for reasons of simplicity.

A robot program is produced by creating a sequence of manipulator end effector positions and other actions, in the workplace. A set of six matrices (for a six link system) represent and define each tool location point in world coordinates, these are saved and written to a run module for execution. (It is proposed that only world location positional data be stored, this will effect a robot independent system) Motion is achieved and illustrated on the graphics terminal by driving the joints from their initial state to the next set (of transforms) via a specified number of intermediate positions. The actual way this is implemented depends upon the control mode in operation, ie, point to point or path control.

During programming, the user supplies sufficient information concerning the end effector (tool point) output state for the structure to be 'solved'. A complete description of the tool point (ie, generalised coordinates) allows the 'coordinate transformation' to be effective. The user defines the tool point position completely and any gripped object orientation is specified with respect to the workplace coordinate system.

The output of the mechanism (or the tool point) is referred to in workplace coordinates through a series of homogeneous transformation matrices. The final frame, located at the tool point, is not a joint but expresses the hand orientated coordinate system in the workplace. If the matrix [G] is the product of these successive transformations (one to represent each joint), then [G] is a global transformation relating the manipulator output (in this case a reference coordinate system situated at the tool point) to the robot base or workplace coordinates.

$$[G] = \left| \begin{array}{c|c} R & \\ \hline & T \end{array} \right|$$

'R' gives the orientation and 'T' the location in base coordinates.

The general relationship for the multi-link mechanism below is:

$$\mathbf{X} = \mathbf{F}(\boldsymbol{\theta}) - \text{eqn. 1a.} \qquad \boldsymbol{\theta} = \mathbf{F}^{-1}(\mathbf{X}) - \text{eqn. 1b.}$$

The output vector 'X' should ideally be of the same dimension as the control vector, in order to solve the structure uniquely.

Equation (1b) is used as a first modelling approach to produce coordinated positional control, using the coordinate transformation routine described below.

A second model, derived from differentiating equation (1), is used for rate control.

$$\dot{\mathbf{X}} = [J(\theta)]\, \dot{\boldsymbol{\theta}} - \text{eqn. 2a.} \qquad \dot{\boldsymbol{\theta}} = [J(\theta)]^{-1}\dot{\mathbf{X}} - \text{eqn. 2b.}$$

This relationship relates the command rate input vector required to control the output of the mechanism along a certain trajectory and with a controlled velocity.

Both modes of control are adopted in the simulation technique by applying position control to create the workplace locations and rate control to specify the required velocities between the programmed locations.

3.1 Coordinate transformations

Coordinate transformations provide solutions to the inverse control problem and enable the end effector of a manipulator to traverse in cartesian coordinates. It provides an expression for the input control vector in order to locate the output of a multi-link mechanism with the desired position and orientation in the workplace. Generally, the inverse control problem is solved by inspection rather than solving the trigonometric polynomials that define the location and orientation of the tool reference point.

Coordinate transformations in the **GRASP** system employ an algorithmic approach where solutions are guided by geometrical inspection. The wrist mechanism is solved first with data provided in the kinematic description file and the arm structure joints then derived. Each particular manipulator structure is analysed in order to transform workplace or tool coordinates into specific joint coordinates.

3.2 Manipulator trajectory and path control

The user specifies the trajectory and rate of motion between programmed points, having control over all manipulator output vector elements. Velocities are associated with each recorded position definition which are specified at the programming stage. Thus the user decides the rates along the three major axes, and rotation rates about these axes. This feature is necessary when, for example, a robot is being used for a seam weld exercise with the gun retaining a constant attitude.

Optional trajectories include point-to-point, straight lines and paths given in terms of mathematical functions.

For controlled path operations the Jacobian matrix (see equation 2), which contains partial derivatives $[X]$, is inverted to produce the required input vector for rate control. The value of $[J]$, depends of course, upon the input vector **X** (the state of the joint angles), and must be calculated for each selected value of X. Path control is implemented by continuous iterations, where the required output trajectory is sampled at discrete intervals.

3.3 Event processor

This provides a mechanism for processing tasks and sub-tasks, containing joint velocities, accelerations and gripper actions, which often occur concurrently, in a way which simulates the action of the parallel processor. The system is analogous to the programmable controller.

A process data structure arranges events (time driven and instantaneous) in a hierarchy and executes them in the correct order. This hierarchy reflects the time driven operation of movements and the order of events as specified under user control. This structuring of processes becomes necessary for multi-robot installations.

The order in which tasks are to be performed is usually determined in the programming stage.

4. Program specification and evaluation

A robot operating system provides user interaction and commands to enable a robot task to be programmed. A task comprises a sequence of manipulator tool locations (and orientations) in the workplace, and other commands such as grips and releases. Movements and tasks are specified at a low level, although it is planned to introduce a high level command structure in the future. This low level form of input is analogous to the teach method, where the tool point (or some end effector reference point) is 'dragged' to specific positions in the workplace.

During the task definition and specification phase the user determines the suitability of the manipulator in terms of reach and geometry, principally by trial and error, thus a working layout is produced by operator design.

Motion is simulated by executing the data contained in the 'run module', and is controlled by the 'event processor'. An estimate of the robot's operation time is computed based upon the manipulator's kinematic performance.

Simulated real time operation is produced by standard computer animation techniques, where a number of camera frames are taken at each time increment.

5. Examples

These examples present typical applications and demonstrate the potential of robot feasibility study by interactive graphics. The work described was performed for industrial companies, and was supported by filmed computer animation illustrating the robot/workplace operation. The graphical output shown here was produced on a drum plotter.

Palletising with an Industrial Robot (Figure 1)

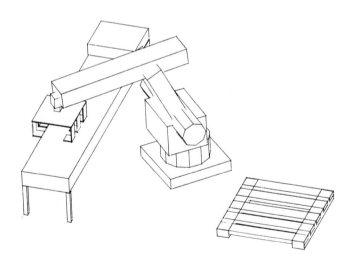

Fig. 1. *Showing input conveyor, industrial robot and pallet base.*

Simulation was used to assess the robot's manipulative capability in terms of reach and geometry to unload bags from an input conveyor and transport them individu-ally onto a base with a palletised packing arrangement. The robot is programmed by leading the manipulator through the required sequence and recording the necessary steps. An approximate task cycle time can be computed based upon the kinematic model.

This example of a multi-robot installation severely stretched the capabilities of the prototype system and led to the need of a more formalised and structured processing facility.

An input conveyor brings the cab into the work-station area and the robots simultaneously weld the various parts of the body. Simulation is used to ensure that layout is such that all welding points can be reached with the appropriate attitude and without fouling, and to examine the overall sequence with a view to planning an optimum process.

Frame handling operation (Figures 2a and 2b)

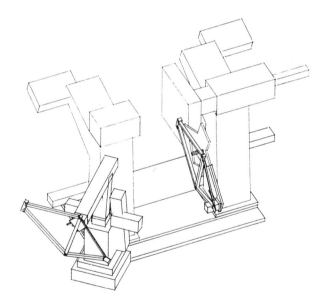

Fig. 2a. *Showing one cell comprising the robot and brazing machine.*

A small industrial robot is required to load and unload frames onto a brazing machine. Simulation is used to check the feasibility of the selected robot for the task and to determine a suitable layout.

A number of alternative layouts were examined and the lower figure illustrates the complete arrangement with the robot requiring to traverse and serve each cell in turn along a powered track. The simulation also investigates problems of interaction and synchronisation between time dependent operations.

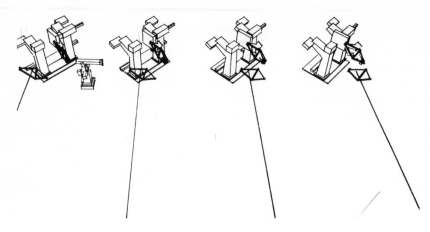

Fig. 2b.   *The overall layout — four cells in line with input rails.*

Automatic welding station (Figure 3)

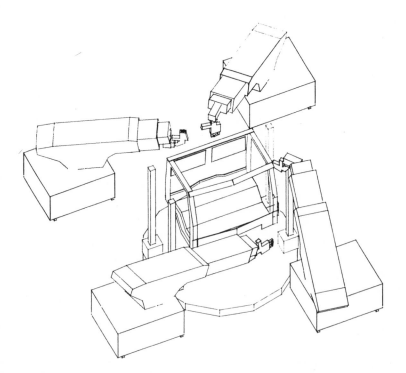

Fig. 3.   *Showing four robots, cab frame and rotary turntable.*

6. Further work

The complete **GRASP** system centres around the ability to produce off-line programming, where robot programs are specified in high level commands and tasks evaluated interactively, with the design being used to provide a task definition for the robot/workplace operation.

Off-line Programming — the inclusion of an interface between the simulator and real robot operation is a natural progression in order to make maximum use of the evaluated information. Initially methods of using the simulation data will be investigated with a specific robot controller.

Collision Avoidance — interference checking routines inform the designer of collision areas, it is proposed to allow the robot to act on this information and perform collision avoidance procedures.

Manipulator Design — graphical simulation is ideally suited for interactive manipulator design. Alternative linkage designs which will be computer generated can be examined for a particular task.

References

BONNEY, M. C., and CASE, K. (1976) The Development of SAMMIE for Computer Aided Work Place and Work Task Design.
6th. Congress of International Ergonomics Association, Maryland, U.S.A. 1976.
HEGINBOTHAM, W. B., DOONER, M. and CASE, K. (1979) Rapid Assessment of Industrial Robot Performance by Interactive Computer Graphics.
9th. International Symposium on Industrial Robots, Washington USA March 1979

Chapter 11

Robot languages–the current position

Elizabeth McLellan

Since their inception in the sixties, all industrial robots have had some means of programming them, which distinguishes them from pure automation.

The methods used on commercially available robots vary from simple 'teach by hand' techniques through to high level programming languages. It should be noted that it is still fairly early days as far as the latter method is concerned.

For most of the application areas in which robots are currently used, the 'teach' methods of programming are adequate and often preferable. However, the increasing use of robots in the area of assembly and within flexible manufacturing systems, has led to the need for both on-line and off-line programming systems.

Before discussing the various programming languages currently in use or under development, a review of various application areas and their associated programming requirements would be beneficial.

As can be seen from Fig. 1, all fields require point to point motion and functional control, though of course the functions will vary from one application to another. Also the majority of areas require three dimensional continuous path motion and the ability to interrogate sensors and make decisions, thus controlling the sequence of operations. An added bonus would be some method of collision checking and avoidance, particularly in the fields of assembly, machine loading and material handling. What then are the implications of these programming requirements?

Programming requirements

In order to illustrate the programming requirements mentioned above consider the simple robot shown in Fig. 2. It can be seen that it has five degrees of freedom, rotation of the pillar with respect to the base, rotation of the arm about a horizontal axis through the top of the pillar, extension and retraction of the arm, rotation of the gripper device about an axis normal to the arm axis and rotation of the gripper device about the arm axis. In the accepted parlance of robotics, the rotary axes may be referred to as the waist, shoulder and wrist joints respectively. Note that there are two orthogonal wrist motions.

	point to point motion	3D continuous path motion	functional commands	sensor interrogation	decision making	collision checking
paint spraying	X	X	X			
spot welding	X		X			
arc welding	X	X	X			
mechanical assembly	X	X	X	X	X	X
electronic assembly	X		X	X	X	
machine loading	X	X	X	X	X	X
material handling	X		X	X	X	X
deburring	X	X	X			

Fig. 1.

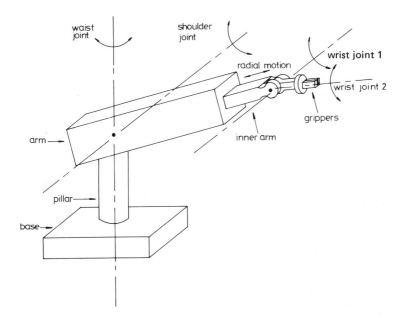

Fig. 2.

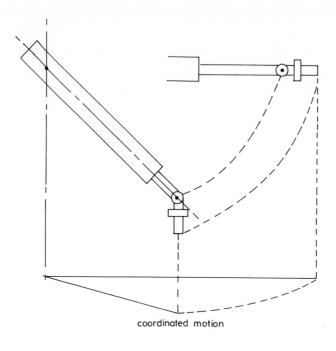

coordinated motion

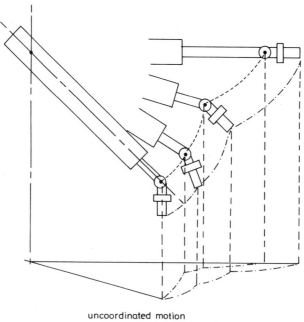

uncoordinated motion

Fig. 3.

Point to point motion

Point to point motion may be co-ordinated or unco-ordinated, in either case the path followed when moving from one position to another is of no significance, only the final location. If motion is co-ordinated all joints will arrive at their final position together, if unco-ordinated this will not necessarily be the case. As a result the path followed by the end of the grippers on our sample robot will be different for these two types of point to point motion, as shown in Fig. 3.

3D Continuous path motion

3D continuous path motion is co-ordinated so that the point of interest, typically the wrist or end of the grippers, follows a specified path through space. This would normally be either in a straight line or along a circular or spiral arc. An example of a straight line, or linearly interpolated move is shown in Fig. 4. The run time computing required to generate this straight line is obviously more complex than that required to achieve co-ordinated point to point motion. Also there is not a unique solution, as can be seen in Fig. 5.

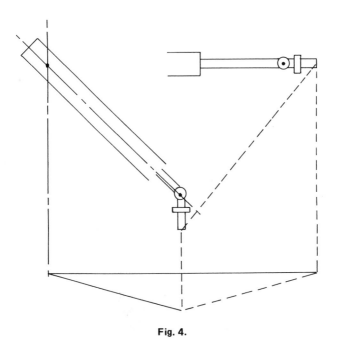

Fig. 4.

A typical application, where continuous path motion is required, would be inserting a peg into a hole. In this case having positioned the grippers so that the peg is aligned with the hole, it is necessary to drive the end of the grippers along a straight line whilst maintaining the alignment as shown in Fig. 6. Therefore as well as extending the arm and rotating about the shoulder, it is necessary to rotate about the horizontal wrist joint.

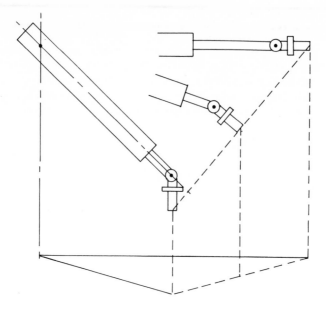

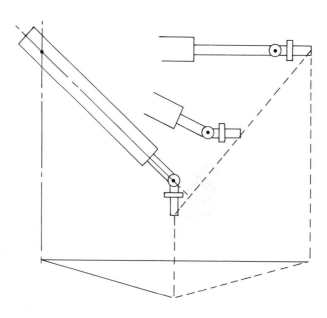

Fig. 5.

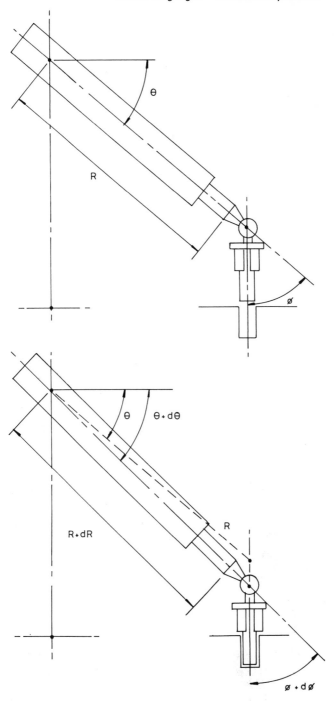

Fig. 6.

Functional control

The functional controls required for particular applications will vary considerably. For example, in paint spraying, control of the nozzle, the rate of flow of paint and the speed of movement of the nozzle through space, might be some of the functions required, whereas for assembly, opening and closing the grippers would be amongst the desired facilities.

Sensor interrogation

As with functional control the sensor interrogation required will vary from application to application. It might be simple binary interrogation of an input channel, indicating that a particular condition exists or not, or it could be detailed interrogation of a visual image to ascertain if the robot is in the required position. Whereas the former type of sensor interrogation is straightforward and very fast, the latter requires powerful computing techniques to achieve the speeds of interrogation necessary to make it worthwhile.

Decision making

Decision making as the result of sensor interrogation is a desirable feature, giving robots some degree of intelligence. The ability to make decisions at run time means that flexibility can be built into the program, allowing the robot to react to varying conditions within a programmed range. Also looping control can be provided within the program.

Collision checking

Collision checking and avoidance is a feature which is highly desirable but at present is not available, although some systems have provision for defining clearance planes, beyond which the motion of the robot need not be carefully controlled. The problems associated with collision checking are considerable, since the programming system needs to know where all objects within the robot's world are positioned with respect to the robot, at all times, and the geometry of all the objects.

User interface

A very important feature of any programming system must be the interface with the user. Ideally this should be easy to learn and simple to use. This is why many of the robot systems in use today are of the 'teach' type. For example, leading the end-effector through the required sequence of movements by hand and continuously recording the joint movements, together with any functional commands required, as typically used for paint spraying robots. Therefore if a robot system is to be programmed using some form of language input, the vocabularly must be appropriate for the environment in which the robot is to be used and the syntax straightforward.

Robot software

Robot software can be divided into two basic categories, on-line and off-line. The on-line software system controls the robot at run time whereas the off-line software generates the instructions required by the on-line software. The off-line system may be closely linked to the on-line system, as in most of the commercially available systems or may be more loosely connected as with many of the systems under development in industry and universities. At present there is no standard interface between the on-line and off-line systems.

Robot languages

A robot language is the user interface with the robot software system. Languages at present available or under development vary from low level languages such as SIGLA[1] which uses elementary instructions to high level languages such as RAPT[2,3] and AUTOPASS[4] which describe operations on objects. A description of some of these languages follows.

SIGLA

SIGLA is the language developed by Olivetti for their SIGMA robot. The main features of SIGLA are: parallel task execution, allowing the simultaneous control of more than one arm operating independently; interpretative structure; and a variable instruction set, which can be tailored to suit the user's requirements, thus maximising the amount of memory available for storing user programs. The instructions consist of two character codes followed by numeric data, of the general form, AA/n1, n2, n3, . . . , which bear little or no resemblance to the command they represent and are thus inherently more difficult to read and write than a higher level language. The basic system runs on a 8 K 16 bit word general purpose mini-computer, having 4 K ROM and 4 K RAM with optional additional RAM available up to a total maximum of 32 K.

HELP

HELP[5] is a high level interactive language developed by Digital Electronic Automation (D.E.A.) and implemented on DEC 11 computers, to facilitate the programming of their measuring machines and more recently their PRAGMA assembly robot. The language has an ALGOL-like block structure, with locations specified in cartesian co-ordinates and actions initiated by calling built in procedures. Other features include looping, branching, subroutines and parallel task execution. An example of the language is shown in Fig. 7.

The HELP system is designed to have the efficiency of a compiler and the interactivity of an interpreter. This is done by phases in which the translator acts as a pure compiler alternating with those which execute the translated program. This happens inside each program element meaning that each single element is translated into a close-to-machine internal language then executed, which allows dialogue with the system during program set up and gives execution efficiency. However, because of the compiler structure of HELP, program execution can be postponed and the translated program can be stored for future recall.

```
WHILE Ø = Ø  00
    MOVE (x50, y20, z100);   : move to clearance above position 1
    MOVE (z80);              : approach position 1
    GRASP;                   : pick up object
    MOVE (z100);             : retract
    MOVE (x100, y50);        : move to clearance above position 2
    MOVE (z40);              : approach position 2
    UNGRASP;                 : release object
    MOVE (z100);             : retract
REP                          : repeat
```

Fig. 7.   *Sample HELP Program*

DEA propose to further develop the HELP language, including the introduction of higher level linguistic constructions to enable more sophisticated process synchronising methods to be set up, and for programming to be carried out in terms of objects rather than end-effector displacements as at present. They also envisage physically separating the compiler from the interpreter structure of the translator, which at present are only logically split, thus releasing more memory for program execution. HELP is implemented on DEC PDP11 range and has been in use at DEA for four years.

VAL

VAL[6] is the computer based control system and language designed for use with Unimation Inc. industrial robots. Initially written in 1975, based on WAVE[7] to run on a PDP11 and rewritten in 1976 for execution on a smaller computer, since when improvements and additions have been made. The present system resides in EPROM of a LSI-11 used to program and control Unimation's PUMA series of robots. It is a highly interactive system, providing many programming aids thus speeding up program development time. The on-line and off-line sections of the system both reside in the same computer, allowing simultaneous control of the robot and interaction with the operator, permitting on-line program generation and modification. The language uses clear, concise and easily understood word and number sequences. It includes facilities such as subroutines, program branching and integer calculations, together with interrogation of and signalling to external devices via an input/output module. Another convenient feature is the ability to use libraries of predefined subtasks. For a VAL program, the locations are taught by moving the manipulator to the required position and recording either its cartesian co-ordinates and the euler angles defining the orientation of the gripper or the joint angles. An example of the VAL language is shown in Fig. 8.

AL

AL[8,9] is a high level manipulator programming language developed at the Stanford Artificial Intelligence Laboratory (SAIL) having an ALGOL-like block structure and syntax. The AL system is comprised of three components: the AL compiler;

```
10 APPROP P1, 20.0  :  approach to within 20 mm of P1
   MOVES P1          :  move (in a straight line) to P1
   CLOSEI            :  close grippers
   DEPARTS 20.0      :  retract by 20 mm
   APPROP P2, 20.0   :  approach to within 20 mm of P2
   MOVES P2          :  move (in a straight line) to P2
   OPENI             :  open grippers
   GO TO 10          :  repeat
```

Fig. 8.   *Sample VAL Program*

```
BEGIN
   FRAME object, new location;
   object ← FRAME (nilrotn, VECTOR (5, 2, 8) * cm);
                                  {defines position of centre of object}
   new_location ← object + VECTOR (5, 3, 4) * cm;
                                  {defines new location for object centre}
   MOVE barm TO bpark WITH DURATION = 5 * sec;
                                  {ensure blue arm in known position}
   OPEN bhand TO 8 * cm;          {open blue hand wide enough to grasp object}
   MOVE barm To object;          {move blue arm to planned initial location
                                   of object}
   CENTER barm;                   {grasp object without moving it}
   MOVE barm TO new_location;     {place object at new location}
   OPEN bhand TO 8 *cm;           {open hand, releasing object}
   MOVE barm TO bpark;            {return arm to park position}
END
```

Fig. 9.   *Sample AL Program*

the interactive AL source code interpreter; and the AL run-time system. The first two run on a PDP-10 under the WAITS operating system whilst the latter runs on a PDP-11 and is written in PDP-11 assembler. The AL run-time system interprets AL pcode which is produced by both the compiler and the source code interpreter.

The AL system is designed to minimise programming time, relying on a symbolic data base and previously defined assembly primitives, together with a quick and simple method of setting up a manipulator program. An example of AL is shown in Fig. 9.

The AL language uses local co-ordinate systems which can be related to each other and has provision for synchronising parallel processes, looping and branching. Also there are instructions for indicating the 'strength of the bond' between adjacent objects.

e.g. AFFIX f1 TO f2 AT t1 RIGIDLY;
      AFFIX f3 TO f4 BY t2 NONRIGIDLY;
      AFFIX f3 TO f4 BY t2 AT t1 NONRIGIDLY;

when AFFIXed RIGIDLY if f1 is moved f2 will move but when AFFIXed NONRIGIDLY, if f3 is moved f4 will remain unchanged whereas if f4 is moved f3 will move with it.

AUTOPASS

AUTOPASS (Automated Parts Assembly System)[4] is an experimental high level programming system for computer controlled mechanical assembly, being developed at IBM Thomas J. Watson Research Centre. The language describes assembly operations rather than manipulator motions, allowing the user to specify an assembly procedure in a similar manner to composing an instruction sheet for manual assembly. Decisions as to the order parts are assembled, the tools used and the general positioning of these objects in the work space are left to the user. The AUTOPASS compiler transforms the assembly procedure specification into a program that directs the robot through the necessary motions to execute the process. The motion commands are generated by using a geometric data base, or world model, to simulate at compilation time the expected run-time world. The geometric data base represents the geometric structure of objects, the spatial positions and relationships amongst objects and the assembly or attachment relationships between objects. A graph structure is used where each vertex represents an object component, object or assembly and may be assigned a symbolic name, the edges are directed and can have four kinds of relationships: part of, attachment, constraint and assembly component. Each object is modelled as a polyhedron, i.e. in terms of their vertices, edges and surfaces, by using a geometric design processor, and accessed by a pointer at the object vertex. These techniques are widely used in the field of computer aided design, and lend themselves to the use of interference algorithms, thus allowing some degree of collision checking to be carried out when computing trajectories.

The compilation process interacts with the user so that any ambiguities detected by the compiler may be resolved. A typical AUTOPASS program is shown in Fig. 10.

RAPT

RAPT[2,3] is a robot assembly language being developed at Edinburgh University. Its syntax is based on that of the APT language which is used worldwide to program numerically controlled machine tools. Like AUTOPASS, RAPT is a world model langauge describing assembly operations in terms of actions upon objects and the spatial relationships between features of those objects, rather than in terms of manipulator movements. The RAPT system sets up kinematic equations for these actions and spatial relationships which are then solved analytically.

```
PLACE servo_ram_main_body IN fixture
    SUCH THAT main_body_base
    CONTACTS fixture_base;

PLACE manifold ON servo_ram_main_body
    SUCH THAT manifold_hole (1)
    IS ALIGNED WITH main_body_tapper_hole (1)
    AND manifold_hole (2)
    IS ALIGNED WITH main_body_tapped_hole (2);

DO i = 1 TO 4;

DRIVE IN m6_cap_screw INTO main_body_tapped_hole (i)
    AT manifold_hole (i)
    SUCH THAT TORQUE IS EQ 12 Nm
    USING m6_cap_screw_driver
    ATTACHING manifold AND servo_ram_main_body;
END;

PLACE control_valve ON manifold
    SUCH THAT control_valve_hole (1)
    IS ALIGNED WITH manifold_tapped_hole (1)
    AND control_valve_hole (2)
    IS ALIGNED WITH manifold_tapped_hole (2);

DO i = 1 TO 4;

DRIVE IN m5_cap_screw INTO manifold_tapped_hole (i)
    AT control_valve_hole (i)
    SUCH THAT TORQUE IS EQ 7 Nm
    USING m5_cap_screw_driver
    ATTACHING control_valve AND manifold.
END;
NAME control_valve manifold servo_ram_main_body
    ASSEMBLY hydraulic_servo_ram;
```

Fig. 10. *Sample Autopass Program*

The APT language has been extended to include means of defining objects and their features as well as actions upon those objects and their spatial relationships.

e.g.　MOVE/B1, PERPTO, (TOP OF B2),−3 would move body B1 towards the top of B2 by 3 units (assuming B1 above B2).

At present the RAPT language does not use the bounded geometry techniques of CAD systems for defining objects, though it is understood that these are being considered.

Conclusions

As can be seen from the foregoing descriptions most of the languages are designed for assembly work and the only ones in use commercially are of the manipulation level. The higher, object or world model level languages are still under development, largely in academic establishments. With the former, the on-line and off-line components of the system can both be run on the same computer, namely the robot controller and are often closely linked. Whereas with object level languages, the off-line or plan-time processing will probably be carried out on a larger more powerful computer, providing an intermediate language which can be input to the on-line system of the robot controller.

For the future it is to be hoped that some standard will emerge for this intermediate language, so that one off-line system can interface with a wide variety of robot controllers and vice versa, as in the case with software for the numerical control of machine tools. Here the interface is a series of cutter locations, together with ancillary commands, known as the CLFILE, which is normally post-processed to produce a control tape for input to the machine tool controller. Such a standard interface will have to be flexible and expandable in order to cater for the variations in application areas and hardware configurations.

References

1. SALMON, M. 'SIGLA – the Olivetti Sigma Robot Programming Language' Proc. 8th ISIR May/June 1978.
2. POPPLESTONE, R. J., AMBLER, A. P. and BELLOS, I. 'RAPT: A language for describing assemblies' The Industrial Robot, Sept. 1978.
3. POPPLESTONE, R.J., AMBLER, A.P. 'An efficient and portable implementation of RAPT' Proc. 1st Int. Conf. on Assembly Automation, March 1980.
4. LIEBERMAN, L. I., WESLEY, M. A. 'AUTOPASS: An automatic programming system for Computer Controlled Mechanical Assembly' IBM Journal of Research and Development Vol. 21, No. 4.
5. DONATO, G., CAMERA, A. 'A high level language for a new multi arm assembly robot' Proc. 1st Int. Conf. on Assembly Automation March, 1980
6. User's Guide to VAL, A robot programming and control system, Unimation Inc.
7. PAUL, R. 'WAVE: A model based language for manipulator control' The Industrial Robot, March 1977
8. MUJTABA, M. S., GOLDMAN, R. 'AL User's Manual' Stanford Artificial Intelligence Laboratory, Stanford Univ. California.
9. MUJTABA, M.S. 'Current Status of the AL Manipulation Programming System' Proc. 10th ISIR, March 1980.

Chapter 12

A language for specifying robot manipulations

R. J. Poplestone
A. P. Ambler

RAPT is a language at what J. C. Latombe (1979) calls the 'object level', that is to say it is a language in which the objects which are to be manipulated by the robot are explicitly represented, and in which the programmer specifies actions during which the robot will cause movement of the bodies to bring about some desired state of the world. As such RAPT should be distinguished from lower level languages such as the VAL language provided with the Unimate Puma robot in which manipulator end effector positions are specified, and from the as-yet-unattained 'objective' level of language in which a desired state of the world is specified and the computational system has to work out the right sequence of actions to attain the desired state. It should however be borne in mind that the use of macros in RAPT can give rise to a 'pseudo objective' level in the sense that a macro can specify a sequence of actions.

User language

For our representation of bodies we have drawn upon the APT language which, while it has a somewhat antique syntax, is widely used for the specification of the behaviour of numerically controlled machine tools. APT allows one to define a range of geometric elements. For instance, if we want to define the body shown in Fig. 1 we will code up a drawing thus.

```
P1  =  point/0,0;
P2  =  point/10,0;
P3  =  point/20,0;
P4  =  point/0,10;
P5  =  point/5,5;
11  =  line/P1,P3;
12  =  line/P2,perpto,11;
.

.
cl  =  circle/center,P1,radius,1;
```

Given this collection of geometric elements we need to specify how they represent features of some body. First we declare which body we are describing

>   body/block;

Next we describe the features, using the geometric elements. E.g.

>   left = face/11,ysmall;

states that the line 11 represents the projection of a face which is perpendicular to the plane of the 'paper'. 'y small' indicates that the direction of the normal to the face is predominantly along the negative y-axis. Likewise the statement

>   step = face/12,xsmall;

defines a vertical face of the step, and the statement

>   hl = hole/cl,zlarge

indicates that cl represents a hole opening out in the positive z direction.
    We are also able to define horizontal faces, for example

> f = face/horiz,2,zlarge

defines a horizontal face at height 2 units.

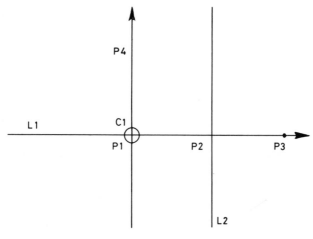

Fig. 1.

The body definition is closed with the word 'terbod'. Features which are defined outside a body definition belong to the 'world', which is a fixed body representing the workplace.

Other means of inputting the necessary body model information into RAPT could be used, for example a graphics system like the Computer Vision system.

Now let us consider the simple assembly shown in Fig. 2.

Here a plate has been placed on the block so that the bottom surface of the plate is against the mating face of the block, the front surface of the plate is against

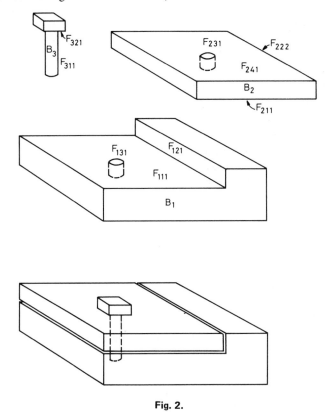

Fig. 2.

the step of the block, and a fastener has been inserted so that it fits the hole in the plate and the hole in the block and pushed home so that the bottom of the head of the fastener is against the top of the plate. Supposing that we have defined the plate and the fastener in an analogous manner to the way we described the block, then the final situation can be expressed in RAPT as:

> against/bottom of plate, mating of block;
> against/front of plate, step of block;
> fits/shaft of fastener, hole of block;
> fits/shaft of fastener, hole of plate;
> against/head of fastener, top of plate;

The general meaning of the against relationship is that the surfaces share a common tangent plane, which encompasses the mating of plane surfaces. The idea conveyed by fits is that of an extended area of contact, although we use against as above for plane surfaces.

Other relationships are available. Thus we can specify that 2 plane faces should be coplanar, and that two holes, or two shafts or shaft and hole should be aligned. When we say that a shaft fits a hole, we understand that the axis directions of the shaft and the hole are opposed. The aligned relation requires that the axis directions be the same. Likewise coplanar faces have parallel normals, whereas faces which are against each other have opposed normals.

The parax relation is used to express the fact that the axes of features of bodies are parallel. It is useful in specifying orientation.

The above statements of course define the objective of the assembly, and RAPT is not powerful enough to work with just them. We have to express the actions which bring about this state of affairs. The powerful inferential system of RAPT permits us to be very parsimonious in the quantitative aspects of the action descriptions, these being mostly derived from the models.

Let us now write the sequence of actions to pick up the plate and put it on the block. (The initial position of these objects will have been described by relational statements similar to those above, expressing how they are placed in the world. We will assume that the plate is positioned right way up).

call/pickup, b = plate, f1 = left of plate, f2 = right of plate,
ax = axis of plate, f3 = bottom of plate;

This is a macro call, whose definition we will discuss later. The effect is that, provided the robot can reach the plate from its initial position by moving first horizontally and then vertically, it will grip the plate by its left and right sides, aligning the axis of the gripper, which is located centrally and parallel to the fingers, with the axis of the hole in the plate, and with the plane of the finger tips coplanar with the bottom of the plate. The macro also asserts that the plate is tied to the gripper, which means that, until they are untied, the system will believe that the plate moves by exactly the same amount as the gripper.

RAPT does not have bounded body models, so it is not possible for the interpreter to check for collisions which may result from a move command.

move/plate, perpto, top of plate, by 10;

This statement causes the plate to be lifted by 10 units, since its initial position is right way up. RAPT has the notion of an agent, which is something which can cause movement. The gripper is part of the robot, which is an agent, and the plate is connected to the gripper by being tied to it, so that the system accepts that the plate can in fact move. Movement can also be transmitted through a sub-assembly, which will be discussed later.

move/plate, parlel, top of plate;

This will move the plate horizontally. The system will infer the extent and direction of the movement by making use of its knowledge of the pickup and put down positions.

```
move/plate, perpto, top of plate;
call/letgo, b = plate;
```

This causes the plate to be moved vertically. If there are no more statements involving the movement of the plate then its position at the end of this statement will be taken to be the same as its final position defined earlier, since the letgo macro unties the plate from the gripper.

Macros perform textual replacement. For example the macro pickup is defined as:

```
pickup = macro/ b, f1, f2, ax, f3;
            move/gripper, parlel, bottom of gripper;
            turn/gripper, about, axis of gripper;
            move/gripper, perpto, bottom of gripper;
            move/lfinger, perpto, gripface of lfinger;
            move/rfinger, perpto, gripface of rfinger;
            against/lfinger, f1;
            against/rfinger, f2;
            coaxial/axis of gripper, ax;
            coplanar/tip of lfinger, f3;
            tied/b, gripper;
    termac;
```

It should be noted that there are two implementations of RAPT, and oddly the second has some restrictions not present in the first, although it makes up for this by being much faster. These restrictions currently include the fact that one cannot specify that a body should be moved with respect to a feature of another body, but only with respect to one of its own features. Also the earlier RAPT would allow one to make statements like:

```
move/lfinger of gripper, perpto, gripface of lfinger, by, n;
move/rfinger of gripper, perpto, gripface of rfinger, by, n;
```

where no other information about the value of n was given, so that the force of the two statements is to cause the fingers to make equal movements, the amount to be inferred from other information.

Subassemblies are a set of bodies between whose features certain specified relationships hold for the duration of the existence of the sub-assembly. Note that the components of a subassembly can move with respect to each other during the existence of the sub assembly, provided that the relations remain valid. For example the gripper can be defined as a two fingers running on parallel shafts:

```
subass/gripper, agent, perm;
fits/shaft1 of gripbody, hole1 of lfinger;
fits/shaft2 of gripbody, hole2 of lfinger;
fits/shaft1 of gripbody, hole1 of rfinger;
fits/shaft2 of gripbody, hole2 of lfinger;
tersub;
```

The gripper is an agent, that is to say it can cause movement. It is permanent, that is to say it exists in every state of the world. Non permanent subassemblies are declared to exist by the statement

     issub/⟨assy⟩;   and   notsub/⟨assy⟩

where ⟨assy⟩ is the name of the subassembly.

Implementation

There have been two implementations of RAPT, which share certain concepts. Both have a set BB of bodies. Each body has a set of axes embedded in it. By a position we mean a mapping from real-3 space to itself which is a product of a translation and a rotation, and may be represented by a 4∗4 matrix.

The statements in the RAPT language specify a distinguished set of states which we will call situations. Situations occur before and after any action. We will think of instances of each body occuring in a situation, so that the RAPT relations holding for a situation are defined between features of the body instances for that situation. The position of a body in a situation is that position which transforms a standard set of world axes so that they coincide with the axes of the body in that situation.

RAPT takes no cognisance of the extent of bodies in space, and consequently the features of bodies can be simply represented. We do in fact associate with each feature its position in body axes, and its type and for features like shafts and holes, a dimension e.g. the radius. Edges are regarded as cylinders of zero radius, and vertices as spheres of zero radius. The position of a feature is chosen so that it maps the body axes according to rules which depend on the type of feature, and we will call these mapped body axes feature axes. The feature axes of a plane face have their Y-Z plane in the plane of the face. The feature axes of a cylinder have their X-axis lying along the axis of the cylinder and the feature axes of a sphere have their origin coincident with the center of the sphere.

The first implementation of RAPT made use of an algebraic processing program. The relations between features of bodies were expressed by rewriting the position of one body as an algebraic expression involving the position of the other. Cycles of relationship gave rise to position equations (just as electrical circuits give rise to equations). The first implementation worked, but was very slow. The second implementation is even now less complete than the first, but works 60 times faster, and it is a detailed description of this implementation which follows.

Encoding relationships

The user specified relationship, against, is recoded depending on the types of surface it holds between. All the spatial relationships are encoded as quintuples

    (r,F1,F2,p,arc)

where r is a relationship symbol drawn form the set relsyms

{fits,agpp,agpc,agcp, agps, agsp,agss,lin,rot,parax,fix,linlin}

The against relation is classified into its various cases, e.g. agpc is used to encode a plane feature F1 being against a cylindrical one F2, and agps a plane against a sphere.

lin rot fix and linlin are relations derived by the interpreter, lin and rot being the basic relationships of mechanisms see Duffy (1980). In F1 and F2 are two edge features, with embedded axes $(X1,Y1,Z1,01)$ and $(X2,Y2,Z2,02)$ respectively, and $(lin,F1,F2,l,sl,arcl) \epsilon$ srel then 01X1 is parallel to 02X2, 02 lies on 01X1 and 01Y1 is parallel to 02Y2, that is to say a lin is equivalent to a prismatic joint which allows one body to move in translation with respect to another.

On the other hand if $(rot,F1,F2,l,sl,arcl) \epsilon$ srel then 01 and 02 coincide, and 01X1 is parallel to 02X2, so that the relationship corresponds to a revolute joint, which permits one body to rotate with respect to the other.

linlin is a combination of two lin relations in sequence, thus constraining two bodies to move in a plane, without rotation.

Finally fix is used to indicate that two bodies have a known relative position in some situation, so that $(fix,F1,F2,l,s4,arc4)$ ($-$ srel implies that $(X1,Y1,Z1,01)$ coincides with $(X2,Y2,Z2,02)$.

p is a parity symbol taking values of 1 and 0, and is used to distinguish between relations which differ only in the orientation of the x-axis of the related features, such as against $(p = 0)$ and coplanar $(p = 1)$. Likewise coaxial and fits are distinguished by having parities of $+ 1$ and 0 respectively.

The fifth element of the quintuple is simply used as a label. Now in RAPT we have a need to make inferences about actions as well as about situations. In order to do this we need to introduce the concept of a body instance, that is to say a pair (b,s) where b is a body and s is a situation. Some kinds of action specification can then be coded as relationships between body instances. For example if we say

move/plate,perpto,top of plate;

then this is equivalent to adding $(lin,F1,F2,p,a)$ to srel, where F1 and F2 are the top of two instances of the plate in situations sl and s2 which are the situations before and after the action.

Manipulations of the relation srel

To explain the RAPT interpreter formally we need to realise that there is not in fact one relation srel, but a sequence of such relations srel(i) holding the current state of knowledge about the task. As the knowledge develops each successive srel(i) will contain more restrictive tuples, lin, rot and ultimately fix will replace fits and against. The most important operations involved in this are (1) a replacement of cycles in the graph corresponding to srel by a single more constrained tuple and

(2) the replacement of body instances which are connected by a fix tuple by a single instance of a 'super body'.

Let us consider for example the case of the block, plate and fastener assembly discussed above. Let b1 be the block, b2 be the plate and b3 be the fastener. Let s1 be the situation in which they are all assembled. Let f111 be the mating face of (b1,s1) – the instance of the block in situation s1, and let f121 be the step face of the block instance. Let f211 and f221 be the bottom face and hole of the plate in s1, and let f311 and f321 be the shaft of the fastener in s1 and the bottom of its head in s1. Then we have the following tuples in srel(1)

agpp	f111	f211	0	a1
agpp	f121	f221	0	a2
fits	f311	f131	0	a3
fits	f311	f231	0	a4
agpp	f321	f241	0	a5

Note that the arc-names, a1 ... will be used to refer to the tuples in srel(i). Fig. 3 shows the graph associated with srel(1). The arcs a1 ... a5 link nodes which are the bodies to which the features referred to in the tuples belong.

The process of deducing the actual positions of bodies in the assembly involves the elimination of cycles in this graph and the addition of extra links. For example the cycle (a1,a2), involving two 'against' relationships between faces of b1 and b2, implies that b1 bears a lin relationship to b2, with movement possible (if at all) along the edge where the faces meet. Thus a1 and a2 can be replaced by a new 5-tuple involving 'lin'.

It is also possible to infer from a3 and a4 that f131 and f231 must be coaxial, giving rise to a new link in the graph.

It should be stressed that there are geometrical computations associated with the above operations. For instance we can only replace the cycle (a1,a2) if the faces concerned are not parallel, and we must compute positions for the new features related by lin. If the faces are in fact parallel, then we can delete one of the arcs since it is redundant.

Cycles of length 2

Many of the problems involved in assembly can be reduced to solving cycles of length 2. Given the fact that relsyms has 12 members we have 144 possible types of 2-cycles, although because some of the relations are symmetric we do not have to consider them all separately. Moreover as we shall explain in detail later, the fix relation if it forms part of a cycle is treated by creating a superbody out of the body-instances involved, maintaining a consistency check.

Table 1 gives the general properties of the essentially different 2-cycles. The 2-cycle is supposed to be derived from 2 5-tuples

(r1,F11,F21,p1,arc1)   (r2,F22,F12,p2,arc2)

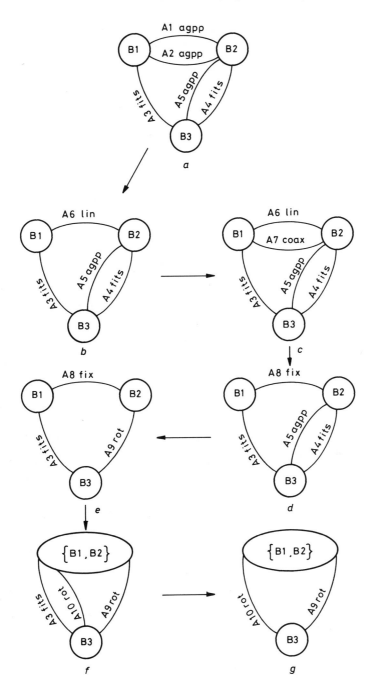

Fig. 3.

Table 1   *Replacing Cycles*

body1 —⟩ F11 —⟩ relation 1 —⟩ F21 —⟩ body2 —⟩ F22 —⟩ relation 2 —⟩ F12 —⟩ body1

Relation 1	Relation 2	General Equivalent	Degenerate cases
fits	fits	fix	lin if F11 parallel F12 fits if F11 colinear F12
fits	agpp	fix	rot if F11 parallel F12 lin if F11 perp to F12
fits	agpc	fix(2)	lin if F21 perp to F22 rot if F21 parallel F22 fits if F11 colinear F12
fits	agps	new	lin if F21 perp to F22 rot if F21 parallel F22 etc
fits	agss	new	rot (2) if origin of F22 on axis of F21 fix if both vertices
fits	lin	fix	lin in F11 parallel F12
fits	linlin	fix	lin if f11 perp to f12
fits	rot	fix	rot if F11 colinear F12
fits	parax	lin	fits if F11 parallel F12
agpp	agpp	lin	agpp if F11 parallel F12
agpp	agpc	lin(2)	lin + rot if F11 parallel F12 agpp if f21 parallel f22
agpp	agps	new	lin + rot if f11 parallel f12 agpp if F21 parallel F22
agpp	lin	fix	lin if F11 perp to F12
agpp	linlin	lin	linlin if f11 parallel f12
agpp	rot	fix	rot if F11 parallel F12
agpp	parax	linlin	agpp if F11 parallel F12
agpc	agpc	new	agpp in certain circumstances
agpc	agpc	new	agpp if F11 parallel F12 etc.
agpc	agps	new	agpp in certain circumstances
agpc	agsp	new	agpp if F11 parallel F12
agpc	agss	new	rot for vertices
agpc	lin	fix	lin if F11 perp to F12
agpc	linlin	lin	linlin if F11 parallel F12
agpc	rot	fix(2)	rot if F11 parallel F12 rot if F21 colinear F22 fix if F11 perp to F12
agpc	parax	linlin(2)	rot if F21 colinear F22 fix if F11 perp to F12
agps	agps	new	etc.

Table 1 *(Cont.)*

body1 $\rightarrow$ F11 $\rightarrow$ relation 1 $\rightarrow$ F21 $\rightarrow$ body2 $\rightarrow$ F22 $\rightarrow$ relation 2 $\rightarrow$ F12 $\rightarrow$ body1

Relation 1	Relation 2	General Equivalent	Degenerate cases
agps	agsp	new	agpc in certain circumstances.
agps	agss	new	rot in certain circumstances
agps	lin	fix	lin if F11 perp to F12
agps	linlin	lin(2)	linlin if F11 parallel F12
agps	rot	fix(2)	rot if F11 parallel F12 rot if center F21 on axis of F22
agps	parax	new	agpp if F11 parallel F12
agss	agss	new	rot for vertices
agss	lin	fix(2)	
agss	linlin	new	
agss	rot	fix(2)	rot in certain circumstances
agss	parax	new	
lin	lin	fix	lin if F11 parallel F12
lin	linlin	fix	lin if F11 perp to F12
lin	rot	fix	
lin	parax	lin	
linlin	rot	fix	
linlin	parax	linlin	
rot	rot	fix	rot if F11 colinear F12
rot	parax	fix	rot if F11 parallel F12
parax	parax	linlin + lin	

and when the features are referred to as being parallel, we mean that the embedded X-axes are parallel. Likewise perpendicularity of features refers to the perpendicularity of the embedded X-axes.

In Table 1 the columns headed 'Relation 1' and 'Relation 2' give the values of r1 and r2 respectively, and the column headed 'General Equivalent' gives the type of relationship that the cycle can be replaced by in the most general case (which is not necessarily the most common). Certain relationships between the features, especially parallelism and perpendicularity, give rise to non-standard cases of the 2-cycle, the more important of which are described in the column 'special cases'. For example, the most general case of the fits-agpp loop actually fixes one body to the other, but in the more common cases either the axes are parallel, in which case we have a rot, or they are perpendicular, in which case we have a lin (see Fig. 4).

In some cases the cycle does not define a relationship between bodies which is a member of the set relsym, which state of affairs is indicated in Table 1 by an entry 'new'. It is highly unlikely that we could find a non-trivial set of spatial relations which would give rise to a closed table. Becasue a 2-cycle is not reducable in Table 1, it does not mean that it will not eventually be reduced, since one of the component arcs may form part of another cycle, which can itself be reduced.

Where '(2)' follows the relation equivalent to the 2-cycle it means that there are two distinct ways in which the cycle can be replaced by a single arc. This has the effect that the sequence srel(i) becomes instead a tree, with the property that a valid solution of either branch is an interpretation of the RAPT program, which is thus ambiguous. In analysing a problem it is wise to avoid replacing such 2-cycles while other transformations are still possible, since the postponed fork may actually disappear, a case of 'why do today what you may not have to do if you put it off till tomorrow'.

In general, replacing a cycle by a single arc will require the computation of new features. This involves computational geometry, most of it straightforward. For example in the case of 2 agpp relationships, which Table 1 assures us can be replaced by a lin arc, we will calculate F13 with axes $(X3,Y3,Z3,03)$ where 03 is an arbitary point on the intersection of the planes F11 and F12, 03X3 lies along the intersection, and 03Y3 lies on the plane of F11. We calculate F23 similarly, and then introduce the tuple (lin,f13,f23,1,arc3) to replace arc1 and arc2.

In the case of relationships with a zero parity it is necessary to multiply the position of the appropriate feature by M, a matrix which inverts the direction of the x-axis.

While the detailed computations involved in Table 1 are beyond the scope of this paper, it is worth noting that for fits-fits cycles, and restricted versions thereof the common perpendicular to the axes of the two features provides a useful basis for the frame of reference of the derived feature.

Putting in extra links to shorten cycles

In our example we have already remarked that because the fastener fits the holes in the plate and in the block we can infer that these holes are coaxial, thus establishing an extra relationship between the plate and block, and turning a 3-cycle fastener-fits-plate-against block-fits-fastener into a 2-cycle plate-coaxial-block-against-plate. This is an instance of a general and valuable inference technique. Suppose we have two tuples (r1,F11,F21,p1,arc1) and (r2,F22,F31,p2,arc2) relating body instance (B1,s1) to body instance (B2,s2), and body instance (B2,s2) to body instance (B3,s3), then we may be able to add a new arc (r3,F12,F32,p3,arc3) where F12 and F32 will in general be new features of (B1,s1) and (B3,s3). For example suppose (b1,s1) is an instance of a table, and (b2,s1) and (b3,s1) are instances of bricks, and the bottom of (b2,s1) is against the top of the table and the bottom of (b3,s1) is against the top of (b2,s1) then we can say that the bottom of (b3,s1) is against a virtual plane parallel to the table top and the thickness of the brick (b2,s1) away from it. Note that if (b1,s1) and (b3,s1) were identical then we would have a degenerate cycle, and this is typical of the cases in which we can insert an extra arc.

Table 2 lists pairs of relationships which allow the insertion of an extra arc under conditions stated in the last column of the table. Since it is desirable not to make the graph bigger than necessary, we only insert new links when there is a prospect of a Table 1 based reduction as a consequence.

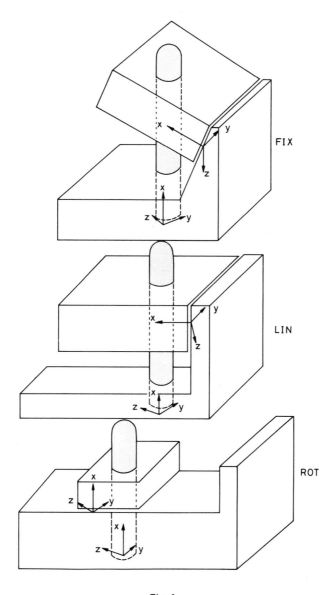

Fig. 4.

Table 2.   *Inserting Extra Arcs*

Relation 1	Relation 2	Equivalent	Condition
fits	fits	fits	F21 same axis as F22.
fits	agpp	agcp	F21 perp to F22
fits	agcp	agcp	F21 same axis as F22
fits	lin	fits	F21 parallel F22
fits	rot	fits	F21 same axis as F22
agpp	agpp	agpp	F12 parallel F22
agpp	agpc	agpc	F12 parallel F22
agpp	agps	agps	F12 parallel F22
agpp	agss	agsp	Both spheres are vertices
agpp	lin	agpp	F21 perp to F22
agpp	linlin	agpp	F21 parallel F22
agpp	rot	agpp	F21 parallel F22
agpc	agss	agps	F22 is vertex on axis of F21, F31 is vertex
agpc	lin	agpc	F22 parallel F21
agpc	rot	agpc	F22 same axis as F21
agcp	agpp	agcp	F21 parallel F22
agcp	lin	agcp	F21 perp to F22
agcp	linlin	agcp	f21 parallel to f22
agcp	rot	agcp	F21 parallel F22
agps	agss	agps	F21, F31 vertices, F21 same origin as F22.
agps	rot	agps	F21 same origin as F22
agsp	lin	agsp	F21 perp to F22
agsp	linlin	agsp	F21 parallel F22
agsp	rot	agsp	F21 parallel F22
agss	agss	agss	all vertices, F21 same origin as F22
agss	rot	agss	all vertices, F21 on axis of F22
lin	lin	lin	F21 parallel F22
lin	linlin	linlin	F21 perp to F22
lin	rot	fits	F21 parallel F22
linlin	linlin	linlin	F21 parallel F22
linlin	rot	agpp	F21 parallel F22
rot	rot	rot	F21 same axis F22

The following cases allow parax to be inferred under weaker conditions.

fits	fits	parax	F21 parallel F22
fits	agpp	parax	F21 parallel F22
fits	lin	parax	F21 parallel F22
fits	linlin	parax	F21 parallel F22

Table 2. *(Cont.)*

Relation 1	Relation 2	Equivalent	Condition
fits	rot	parax	F21 parallel F22
agpp	lin	parax	F21 parallel F22
agpp	linlin	parax	F21 parallel F22
lin	lin	parax	Under all conditions
lin	linlin	parax	Under all conditions
lin	rot	parax	Under all conditions
linlin	linlin	parax	Under all conditions
linlin	rot	parax	Under all conditions

Completing the example

Fig. (a)–(g) shows graphs corresponding to the sre1(i). We are now in a position to complete the example illustrated in Fig. 3. From the tuples corresponding to a1 and a2, by applying the agpp-agpp line of Table 1 we obtain sre1(2) by adding

      lin   f141   f251   1   a6

to sre1(1) and deleting a1 and a2. Next by making use of the fits-fits line of Table 2 we infer the alignment of the holes

      fits   f131   f231   1   a7

and add this tuple to sre1(2) obtaining sre1(3). Next a6 and a7 can be combined using the fits-lin of Table 1 to give

      fix   f151   f261   1   a8

By combining a4 and a5, using the line fits-agpp of Table 1, we get

      rot   f271   f331   1   a9

Given that we have inferred the fix relationship we form a new sre1(6) in which the plate and block are combined. An anologous process is described in Mazer (1981). We choose one of the body axis sets to be the axis-set for the merged bodies, and rewrite all of the feature positions of the other body(ies) in terms of the new axis set, using the feature positions contained in the specification of the fix to perform the relocation. The advantage of performing this merge operation is that the graph is simplified. This is extremely important because any set of bodies which are fixed form a clique which is rich in cycles of trivial importance. When such a merge is performed it is necessary to verify that any other relationships which should hold between features of the merged body-instances do in fact hold.

Ties

When one body is tied to another for the course of an action the two must make the same movement. This means that any relationships which hold between the bodies before the action must hold after, and conversely. Moreover any descriptions of the motion of one must apply to the other. These rules are readily implemented by examining srel(i) and propagating the relations as described. Care is needed when an instance of a tied body is merged. If bodies in srel(i) are merged, giving srel(i + 1) where all references to the unmerged body instances have been deleted, we cannot do the same thing with those body-instances in the tied relation, since the tying refers to the body instances, and is not inherited by the merged bodies. Just because I know where the plate is with respect to the block in situation s1 and so their instances can be merged, it does not mean that if it is tied to the gripper during the following action that the block is also tied.

Subassemblies

The easy way of treating subassemblies would be to expand the set of relationships which hold in the sub-assembly in every situation in which the sub-assembly exists. This would in general involve duplication of work, since any cycles occuring in the sub-assembly would have to be reduced in each copy of it. To avoid this problem we first process each sub-assembly in a 'general situation', s0, which has no associated actions, and then transcribe the resulting relationships into each situation in which the sub-assembly exists.

Working out which bodies move

Bodies are assumed not to move unless they are connected to an agent. Two bodies are connected during an action if either they are tied to each other before the action or they both belong to a subassembly which exists in the situation immediately preceding the action, or if they are connected to bodies which are themselves connected. The system represents the non- movement of bodies by inserting fix nodes connecting their instances in the situations before and after an action.

Other APT based robot work

I understand that an APT-based robot language has been developed at Mc.Donnel Douglas, but details are not available outside of the United States.

References

DUFFY, J. (1980) Analysis of Mechanisms and Robot Manipulators, Edward Arnold, London.
LATCOMBE, J. C. (1979), Methodes et langages de programmation des robots industriels, Seminaire International, IRIA.
MAZER, E. (1981) Realisation d'un support experimental de recherche pour le projet Pandore:

Definition et implemenation du langage LM. These de 3eme cycle, Institut National Polytechnique de Grenoble.

POPPLESTONE, R. J., AMBLER, A. P., and BELLOS, I. M., (1979), An interpreter for a language for describing assemblies. Artificial Intelligence 14, North Holland Publishing Co., Amsterdam.

Acknowledgements

Table 1 has been corrected and extended by making use of the work of Dr Tamio Arai of Tokyo University.

We thank the Science Research Council for support.

Chapter 13

Linking industrial robots and machine tools

M. Edkins

1. Introduction

The industrial robot shares a lot of its technology with the numerically controlled machine tool. In both cases it is the programmability of the machine that allows sophisticated control. The linking of both types of machine can produce a very flexible system capable of manufacturing a variety of metal parts. This paper is intended to give a brief comparison with other systems and to explain how the potential of robot based systems may be exploited and some of the problems involved.

2. Existing automated manufacturing systems

Before considering in more detail the interaction between robot and machine tool it will be useful to consider some of the various approaches to automating the manufacturing of machined parts as this illustrates some of the strengths and weaknesses of using industrial robots. Note that all of the systems described are based on a Group Technology (GT) approach, that is instead of the traditional method of grouping similar machines and transporting the part between them, a number of different machines capable of performing all the required operations are grouped together.

2.1 Transfer lines
In a transfer line a number of special purpose machine tools are linked by a transport mechanism. A product change on such a system may involve a major redesign of the system and so it is only used in high-volume applications.

2.2 Flexible manufacturing systems
By replacing special purpose machinery with Numerically Controlled (N.C.) machine tools, a different component may be manufactured by simply altering the N.C. programs.

The components are mounted on pallets, and to allow different routing between machine tools a number of different transport systems can be used.

(1) A circulating conveyer system.   (see Fig. 1)
(2) A stacker crane   (see Fig. 2)
(3) Robot trucks
(4) Robot arm   (see Fig. 3)

A supervisory computer is required to track and schedule the transport of pallets.

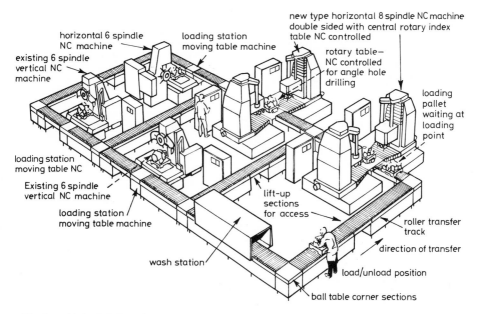

Fig. 1.   *N.C. link-line for machining various Borg-Warner automatic transmission cases. (Installed about 1969)*

Such systems are particularly suited to the manufacture of large complex components. For smaller systems the cost of pallets and the machine tools to handle them can be prohibitive. Also such a system may be difficult to run manually in the event of a serious malfunction in the transport system.

An attractive alternative is to adapt existing N.C. machine tools so that they may be loaded with a workpiece directly by a robot arm thus removing the need for pallets. By using the machine in a manner closer to that of the manual operation it was designed for, a lower cost system may be produced which may be run manually in the event of a robot's breakdown. An example of this type of system is shown in Fig. 4.

In the next sections, the requirements of the robot and the machine tools will be considered.

3. Robot requirements

Consider again the layout in Fig. 4. Note that there are a limited number of points in the cell that need to be defined and the path between these points is not critical. By suitable design of gripper these points need not be specified to a high degree of

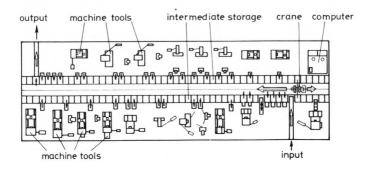

Fig. 2.   *Stacker crane type flexible manufacturing system*

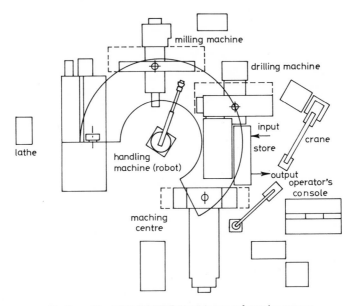

Fig. 3.   *The NTH-SINTEF flexible manufacturing system*

accuracy. An example of good gripper design is the FANUC hand see Fig. 5. On pick-up the wide throat and self-centring action of the 3 fingers will accommodate an out of position work-piece. Similarly when loading a chuck, the compliance in the hand will allow the piece to move into alignment with the chuck. Thus it is the chuck that controls the positioning accuracy, rather than the robot.

Initial results from simulation of the behaviour of the cell under construction seem to show a low sensitivity for production rate against robot speed, where components with machining times greater than five minutes are being manufactured.

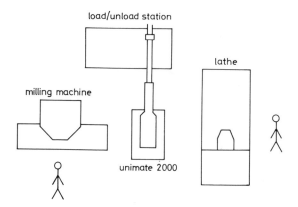

Fig. 4.   *U.M.I.S.T. F.M.S. Cell Layout*

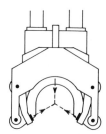

Fig. 5.   *A pneumatic gripper designed for handling turned parts ranging from 20 to 210 mm diameter, and up to 15 kg weight. The three fingers give a self-centring action on pick-up. On loading a chuck the built-in compliance in the gripper ensures that the chuck and not the robot controls the position accuracy.*

The robot requires a lot of information about its environment if it is to run reliably without manual supervision. At its simplest it means a simple handshake interlock for every transaction e.g. 'open the safety guard' – 'the guard is open' signals. This can soon exhaust the number of input/output lines on a standard robot. In our case we are using the standard sets of 6 inputs and outputs and decoding them as binary numbers in the range 0–63 using external circuitry. This allows a large number of switches to be used but restricts the robot to reading or setting only one switch at a time. An attractive alternative to a large number of microswitches would be a low resolution camera and image analysis system to give the appropriate information.

As the robot is involved in a large number of simple transactions each with possible error conditions there is a need for a conditional branch out of sequence. Testing and maintaining the robot programs, may give problems similar to those for conventional computer programs containing a large number of GOTO statements. In both cases the provision of more complex control structure, particularly subroutines, can greatly simplify problems.

To summarise, for a robot in a machine tool loading application the requirements for the mechanical drive systems speed and accuracy need not be stringent. However the controller is required to be able to cope with large programs of complex structure and with a large amount of interaction with the outside world.

4. N.C. Machine tool requirements

The first Numerically Controlled (N.C.) machine tools were first implemented in the 50's and 60's by hard-wired logic. These are now being replaced by Computer Numerically Controlled (C.N.C.) systems which are based on a mini or microcomputer, see Fig. 6. The processing power in these machines allows storage and editing of programs, connection to a remote computer, self diagnosis and comprehensive error detection.

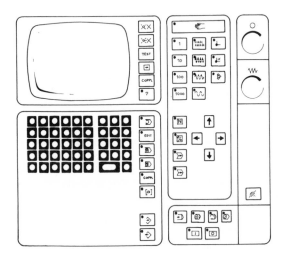

Fig. 6.   *Bosch CNC micro 5*

However, most of the commercially available controllers do not as yet have facilities for interacting with a robot or for unmanned operation. The program controlling the axis motions is expressed as numbers on a punched tape e.g. 'X46752Y100672', see Fig. 7. Additional functions may be programmed using M & G codes e.g. program stop, spindle speed, tool change, coolant on/off etc. A proposal is currently under discussion to add to these standard codes.

Allow tape leader 1½ metres (5 ft approximately) of blank tape

N3	G 2	X ± 43	Y ± 43	Z ± 43	I 43	J 43	K 43	F 43	M 2	End of block
Seq. No.	Prep. func.	X ordinate	Y ordinate	Z ordinate	I valve	J valve	K valve	V/D feedrate	Misc. func.	EoB
o/o EoB										
N001	G92	X0	Y0	Z0				F0		EoB
N002	G90							F0		
N003	G01		Y-87700					F0		
N004				Z-22000				F0	M03	
N005				Z-30000				F12500		
N006			Y-60000					F3610		
N007			Y-82700					F8810		
N008		X35000						F2857		
N009	G03	X62482	Y-58472		I0	J27700		F3610		
N010	G01	X67521	Y-20274					F2595		
N011	G03	X67700	Y-17429		I22506	J2845		F4405		
N012		X52296	Y4066		I22700	J0		F4405		
N013		X0	Y-12700		I52296	J154066		F615		
N014		X-52296	Y4066		I0	J162700		F615		
N015		X-67700	Y-17429		I7296	J21495		F4405		
N016		X-67521	Y-20274		I22700	J0		F4405		
N017	G01	X-62482	Y-58472					F2595		
N018	G03	X-35000	Y-82700		I27482	J3472		F3610		
N019		X-9409	Y-65600		I0	J27700		F3610		
N020	G01	X-2125	Y-48017					F5334		

Fig. 7. *Hayes Tapemaster Milling Machine – Manual Part Programming Sheet*

Where a C.N.C. system has an interface allowing connection to a remote computer this is usually only used for downloading N.C. programs thus avoiding the use of paper tape. Some also allow the machines status to be read by the remote computer.

To permit unmanned operation it will be necessary to extend this to allow commands such as cycle start, set datum and feedrate over-ride to be controlled from a central computer.

In addition there will be a need for automatic swarf disposal and cleaning, automatic clamping and tool breakage detection. All of these tasks are normally undertaken by the human operator and are surprisingly difficult to perform automatically with the required level of reliability.

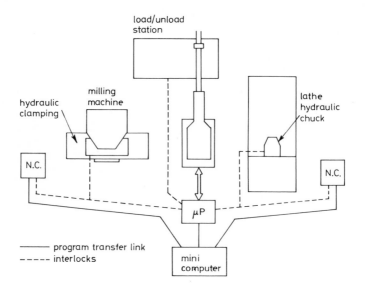

Fig. 8.   *Machine Cell Interconnections*

5. The interlinked system

The basic elements in a robot based machine tool cell have now been outlined, so consider now their interlinking as shown in Fig. 8. A central computer holds all of the machine tool and robot programs on disc files. When a workpiece arrives it is identified. Scheduling software then checks whether the workpiece can be loaded onto a machine tool or needs to be stacked in a storage area. The appropriate robot and machine tool programs are down loaded and activated. Alternatively if the controllers had sufficient storage to hold all the appropriate programs the central computer need only define which program is to be run. A similar scheduling and

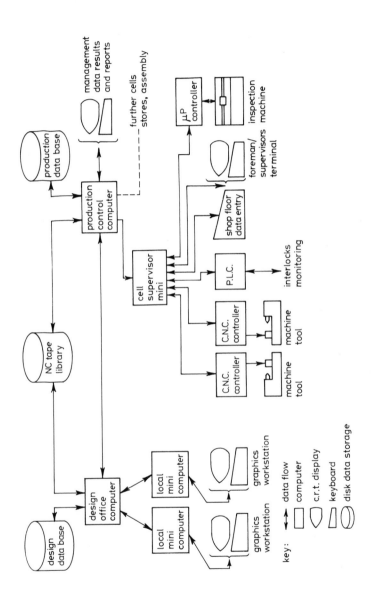

Fig. 9. *Hierarchical distributed computing CAD/CAM system*

action sequence would occur for every event in the cell e.g. workpiece machining finished, machine breakdown etc.

An important role for the central computer is that of supervision during unmanned operation. A comparatively low resolution vision system to compare the actual and programmed movements of all of the machines would be very useful in detecting catastrophic faults.

In a completely integrated computer aided manufacturing (CAM) system a cell of this form would be one of many and these would be linked together via the production control computer. This in turn would be linked to a computer aided design (CAD) computer in the design office (see Fig. 9). Once the geometry of a component has been specified the N.C. program for this component can be generated automatically. Similarly with the aid of software like the SAMMIE system, the robot program could be written off-line and proved with the aid of a graphics display. In such a system the process of getting from the design to the finished product would be almost completely automatic.

6. Conclusion

This short talk has attempted to outline the alternative ways of automating production and show the advantages of a robot based system. The characteristics required by the robot and machine tools were then discussed, and how these could be linked together to form a manufacturing cell, and in turn how these cells could be linked to form an automated manufacturing facility.

Chapter 14

The design of industrial robot manipulators

B. L. Davies

1. Introduction

The use of robots in industry has increased in recent years as the benefits have become apparent. Whilst they are obviously helpful in replacing human labour in dirty or dangerous tasks, they also have an articulated structure which permits access to areas, such as inside the framework of a car, which would otherwise require considerable complexity in a more conventional machine-tool type of structure. This tends to make robots more reliable than special purpose automation. A further advantage is the repeatability of performance compared with that of a human operator which results in less scrap or reworking of the product. It is only recently that the more frequently claimed advantage of reprogrammability for use in small batch production, has in fact been widely utilised by industry due to the improvements in the previously lengthy and tedious process of programming the robot for sophisticated activities. Improvements in programming facilities and in sensing systems have recently resulted in demands for improved speed and load capabilities in tasks such as assembly or in the semi-disordered environment, where the exact nature and location of an object being presented to the robot may not be precisely defined. This need for increased speed and load capability has caused difficulties in the power and control of robot manipulators. In spite of the great advances in control theory, computing and electronics, the mechanical systems of robots tend to be underdeveloped and have changed relatively little since the 1950's. Whilst the mechanisms have been adequate up to now, the increased demands on robot performance mean that the mechanical systems are becoming more and more unsatisfactory. These mechanical problems will limit the potential application of robots because speed of response, accuracy, load capability and sensitivity to force or touch are all fundamentally restricted by the design of mechanical structures, power system and power transmission devices. Thus, no matter how sophisticated the computing and control strategies, if the device is under powered or overweight having structures and power transmissions which have excessive backlash, friction or flexibility, then the robot will be fundamentally limited in its capability.

2. Power systems

The nature of a robot manipulator structure gives rise to very adverse loading of the actuation device. In order to allow access to the inside of structures in tasks such as spot welding and forge unloading, it is necessary to have an open kinematic chain in which one end of the structure is fixed to a sturdy base and the free end is cantilevered via a series of prismatic or revolute joints. A compromise is often adopted in which the most highly loaded joints at the base are fixed to an overhead gantry type of structure, usually moving in XY coordinates, which can be supported at each end. For the more highly loaded movements, this gives increased rigidity and the ability to support heavy actuators whilst not greatly inhibiting the ability to reach inside structures. To avoid the manipulator getting in the way of larger workpieces, its profile needs to be as slim as possible and so the size of joints and structure is kept at a minimum. This combination of a long reach cantilevered arm with a small radius of action at the joint gives rise to very adverse loading, particularly the inertial loading reflected at the actuator, since the inertia increases as the square of the radius. The narrow structure gives rise to increased flexibility which can adversely affect stability.

Hydraulic power, which has good power to weight and power to volume ratios, is an ideal medium for the adversely loaded manipulator. The actuators can be made sufficiently small and light to fit in a narrow arm structure, whilst the supply pressure can be raised to give an adequate stall force. Nevertheless the ratio of applied load to stall load is usually much higher than is found in machine tools. One disadvantage of hydraulics is in the cost, not only of the hydraulic power supply but also of the electro-hydraulic control valves. A general manipulator requires 6 separate joints in order to reach any position in the work volume, 3 to position the gripper in space and 3 to orientate the gripper at the required angle. Many commercial manipulators adopt only 5 motions, by pointing the manipulator directly at a restricted working area. A separate valve is required for each motion and to achieve the desired response times, the valve must be of the high quality torque motor driven, 2 stage, type. Some work is in progress to produce an adequate response from cheaper solenoid valves but so far this has met with little success.

Controllers for very high power electric servo-motors are also extremely costly. Electric motors suffer from the further disadvantage that if the heavy motor is mounted directly on the arm, the payload is decreased. If the motor is mounted on an external structure, however, there is a need for a power transmission system that will inevitably introduce backlash and time delays. Nevertheless, many manipulators use electric power and some of the larger models, such as the ASEA welding robot, mount the larger motors for the heavily loaded joints on the base pedestal with a rod type of power transmission, whilst the wrist motors for rotation and bend are mounted direct on the wrist. As a compromise of both cost and function there is a lot to be said for using a hybrid system of electric power for the wrist motions and hydraulic power for the more heavily loaded joints that are used to position the wrist in space.

3. Research at UCL.

Robotics work at University College London[1] has been concerned with attempts to gain information from the semi-disordered environment, such as that in a factory where little or no attempt has been made to orientate or position components on a conveyor, or in natural surroundings which employ even less order e.g. in a remote underwater vehicle trying to lay a pipeline. The philosophy which we feel will yield best results in future advanced automata is one in which the manipulator is used to reach out into the environment to interact with objects. The interaction results in some disturbance in the manipulator or in the object itself, however slight, which can be sensed by more than 1 stream of data, e.g. from position or force sensors or from video camera. The correlation of these streams of data yields information about the object leading to further strategies of interaction and sensing which can eventually lead to a very good concept of the object. Since the information is found by direct mechanical interaction, it is of direct relevance to the process of manipulation and avoids the confusion arising from painted stripes and shadows which is present in the more usual, purely visual, system. Current systems generally use the manipulator only as a means of executing strategies which have been decided upon by analysis of the visual scene. This visual scene-analysis approach usually requires considerable pre-programming to cover all the possible permutations and combination of the situation. The difficulty with mechanically gathered information is it is only as good as the manipulator which was used to obtain it. The presence of backlash and high inertia of moving parts can degrade the quality of information obtained. Present day manipulators tend to be very poor in their mechanical design and so it was felt that a new type of mechanical design was required that minimised the non-repeatable type of inaccuracies without deterring from the dynamic response that is required for low transition times. The UCL manipulator indicates one approach to this problem.

4. The U.C.L. manipulator

The manipulator shown in Fig. 1 was designed to have a good dynamic response in both position and force mode. Backlash has been eliminated by the use of push-push actuators. The first two joints form the equivalent of a shoulder elevation and rotation, whilst the third joint is extended to form the equivalent of an elbow rotation about a vertical axis giving a 'reach' capability. The 'shoulder' castings are coupled directly together to give a considerable rigidity and provide the facility for a large output torque in a small space making it ideally suited to orientation tasks. The overall weight has been minimised by using sculpted aluminium alloy castings which were machined on a N/c milling machine. The castings contain loosely fitting steel liners into which fit honed pistons that have contact along their complete length. This gives very good wear properties and, with the addition of piston seals, ensures a minimum of leakage. The ends of the pistons in the first casting rest against hardened steel rollers mounted on the second casting so that as the second casting rotates, the rollers ride across the ends of the pistons. This avoids the need for precise location of the first casting with reference to the second. Thus apart

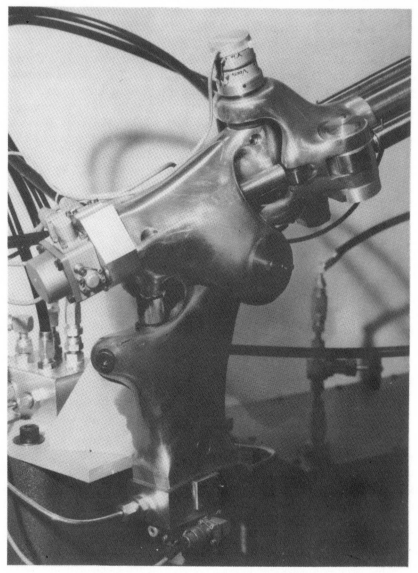

Fig. 1.   *View of electro-hydraulic manipulator*

from the honed fit of the piston in the liner, tolerances on the components can be relatively wide and the cost of manufacture thereby reduced. Each piston pushes either side of the subsequent casting axis and this results in virtually no backlash.

A 'Moog' series 77 electrohydraulic valve is placed on the ends of the actuators and is ported directly into the linear, via a manifold to give minimum transmission delays and reduce the need for flexible hose connections. Each joint is designed to carry both position and pressure transducers. The principle of operation for the second motion is similar to that of the first. However the third joint is extended to give the equivalent of an elbow rotation about a vertical axis. This design is a compromise to give adequate reach for a lightly loaded terminal device suited to the research tasks being undertaken. As a result the third motion has affected the rigidity inherent in the previous two joints. The system actuation consists of long push rods, attached to the pistons, and placed either side of a central support tube which supports the 'elbow' joint. The manipulator is controlled digitally using a DEC. PDP 11/10 minicomputer as part of the feedback loop. A further 2 models of this arm have been constructed for other UK research centres. An additional 3 motions are planned for the wrist joints but it is anticipated that these will be electric motor driven.

5. Control of the manipulator

The constraints described above imply certain difficulties in the control of robot manipulators. They tend to have very high ratios of applied to stall force, high inertial loading, structural flexibility together with large values of coulomb friction and striction. These difficulties are generally ignored in the more recent simulations on hydraulic robotic systems in which the force and flow equations have been linearised so that they have little relationship with reality and the conclusions drawn can be misleading. An example of this is in the use of acceleration feedback. In the U.C.L. manipulator acceleration feedback in small quantities was found to be stabilising as is predicted by the application of Routh stability criteria to the linearised closed loop equations. However, larger values of acceleration feedback were found in practice to make the system unstable at the same loop gain values. A digital simulation of the system, including non-linear effects, has indicated that this is due to the relatively high striction inherent in this manipulator design.

Further problems occur if one is attempting computer control of the torque to give an optimum motion to the manipulator, because the torque is a function of the overall kinematics. The torque will vary, therefore, not only as a larger or smaller load is held in the gripper but also as the inertia changes due to the load being rotated at greater or smaller distances from the shoulder. The manipulator kinematics cause further difficulties. It is relatively easy to predict the motion of the gripper given the motion of each individual joint. However, the converse prediction of the motion of each joint in order to achieve a desired motion of the gripper is extremely lengthy to compute. The result is that the kinematic design of the manipulator is often tailored to allow the dynamic equations to be solved in a reasonable time to permit direct computer control of the robot. For example, from work in kinematics it is known that if 3 of the 6 axes of motion in a manipulator intersect at a point then it is always possible to find a solution to the kinematic equations, no matter what configuration is adopted for the manipulator. For most conventional manipulators the effect of 3 intersecting axes, usually at the wrist, is to make the system equations much simpler so that a change in orientation has little or no effect on a change in position and vice versa.

Much work has been carried out by kinematicians to solve kinematic equations for the most general form of manipulator design.[2] Great emphasis is placed on the ability to have closed-form kinematic equations, rather than iterative solutions. However, although closed form equations give a number of solutions which may be of benefit in obstacle avoidance, they are in practice going to be used in computer control of the dynamics of the manipulator. Closed form equations will then appear in the computer as small step increments and there is often little to choose between these methods, in practice, for dynamic control.

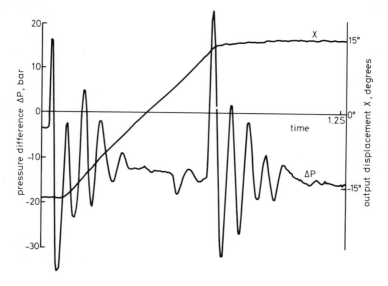

Fig. 2.   *Response to ramp displacement of second degree of freedom*

To minimise inertial shocks and aid stability at high loop gain, some of the more advanced manipulators adopt signal shaping. Figure 2 shows the pressure and displacement against time for a single joint on the UCL manipulator undergoing a normal ramp displacement. The large pressure fluctuations at the beginning and the end of the stroke give an indication of the inertial shocks seen by the system. Figure 3 shows the same magnitude displacement taking place over a similar time. Here the computer is used to maintain the acceleration constant up to halfway and then switch to a constant deceleration. The reduction in magnitude of pressure fluctuations in Fig. 3 gives an indication of the reduction in inertial shock felt by the manipulator and shows this to be a safer strategy for the transport of fragile items held in the gripper. Similar strategies involving a trapezoidal acceleration profile have also been adopted to further minimise pressure pulses seen when switching from acceleration to deceleration.

6. Force control

In addition to the more usual position and velocity control it is often useful to operate the manipulator under closed loop force control so that a known force can be exerted against the external world. This force can be used to gain an impression of the forces during assembly, the weight of a component or even the characteristics of an object e.g. if it is 'springy', has high friction, etc.

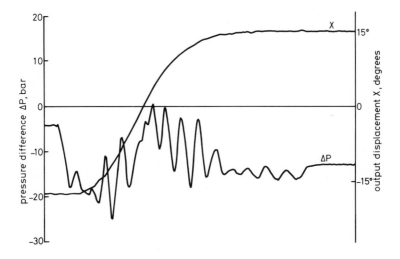

Fig. 3. *Response to constant acceleration/deceleration displacement of second degree of freedom*

The usual way that force information is obtained in hydraulic systems is to look at the actuator pressure difference. In the manipulator, however, the actuator is often located at a distance removed from the gripper by a series of links and joints. The flexibility, inertia and friction of the joints, links and seals gives only an approximate indication of the force at the gripper. Electrically powered systems can have an indication of the force applied by measuring motor current. Here too the current is only a rough guide to the force at the manipulator tip, and is confused by friction and backlash in the motor gearbox and joints as well as flexibility and dynamic forces due to the acceleration of the various inertias. For these reasons, some type of force sensor located at the gripper is advantageous. For specific tasks e.g. grinding, a single or bidirectional force transducer is often sufficient and can be made robust and small. To give full force information in the general situation, however, 3 forces and 3 moments are required. Traditional force/moment transducers have been available until now only in robotic research establishments and tend to be either bulky or non-robust, with a limited dynamic range. They also suffer from the problem that in conjunction with arms with a poor dynamic response, the total cycle time to sense a force, and take the arm through a computed compensating motion to minimise the force, is excessive.

Fig. 4.   *Prototype 6 degree of freedom force resolver*

An attempt to provide a sensitive fast acting robust force/moment transducer that can be placed at the 'wrist' is shown in Fig. 4 which shows a prototype model based on the type of articulation found in stabilised radar platforms or aircraft simulators.

The particular problems of close insertion tasks, e.g. pushing a bearing in a housing, has lead to the development of virtually a special tool in the form of the remote centre compliance device developed by Draper Labs in the USA.[3] By using a spring compliance, centred about the object being inserted, close insertion tasks can be undertaken very quickly. However the device has been criticized as not robust enough for industrial needs and suffers from the problem that if the device is left in a compliant state during transition, when the destination is acquired the gripper and load oscillate for some time and only when they are stationary can insertion begin. These oscillations can be avoided by using a separate clamping action during transition, but this means forces can only be registered after unlocking. For general tasks these limitations mean that further force resolver development work is necessary.

The use of hydraulic actuator pressure difference, as a rough measure of acceleration, is often advocated for stabilisation feedback purposes. However, its use in manipulator control should be adopted with care as it can, like acceleration feedback, result in a de-stabilising effect when used in large quantities.

7. Conclusions

The industrial robot manipulator presents many challenges in the design of hardware and control systems. If care and ingenuity is adopted in the mechanical system design to minimise inertia, backlash, friction and flexibility then the structure, power system and transmission will be better able to utilise the improvements currently available from computing, electronics and sensing systems. Only then will we be able to have really fast, precise, robots that are sufficiently sensitive to interact with their environment.

References

1. DAVIES, B. L. and IHNATOWICZ, E. 'A computer controlled manipulator system for deriving the mechanical characteristics of objects'. Proceedings of 2nd IFAC symposium on problems of information and control in manufacturing technology. Stuttgart, Oct. 1979.
2. DUFFY, J. 'Analysis of mechanisms and robot manipulators'. Pub. E. Arnold, 1980.
3. WHITNEY, D. E. & NEVINS, J. L. 'What is the remote centre compliance and what can it do'. Proc. 9th I.S.I.R. Washington 1979,

Chapter 15

Open-die forging operations by industrial robot

E. Appleton

Introduction

The work described here was undertaken at the University of Nottingham by staff, all of whom have since moved to PERA or the University of Cambridge. The principal investigators were Professor W. B. Heginbotham of PERA, Dr E. Appleton of the University of Cambridge, and Dr A. K. Sengupta who recently resigned his post at Cambridge to return to the Centre for Iron and Steel, Steel Authority of India Ltd.

The work can be divided into two aspects.

(i) The detailed investigation of bar forging and the relating of robot program derivation to fundamental deformation studies.

(ii) An engineering exercise to study the feasibility of use of an industrial robot for ring forging.

The combination of noise, dirt, air pollution, heat and the potential dangers inherent in deforming sizeable pieces of hot metal in a Forge Shop makes forging one of the most unpleasant and hazardous jobs in industry. This, coupled with the rising cost of skilled labour, makes the forging operation an ideal candidate for application of automation.

Application of industrial robots in the forging industry has been mostly restricted to closed-die-forging with 'pick-and-place' type operations for transferring workpieces from furnace to press, from one die cavity to the next and so on. In open-die forging, where complex manipulative movements and frequent changes of their sequence are necessary, the capabilities of an industrial robot appear not to have been fully expoited.

Open-die forging of complex shapes of light workpieces calls for numerous manipulative actions on the part of a skilled operator and modern industrial robots should be highly qualified to take his place. The manipulative actions required include picking up, loading in the press, in-press movements and unloading. In addition it may be necessary to remove the workpiece from the press, and reorientate the workpiece in order to forge the previously gripped portion. It should be possible to pre-programme a robot, either in 'Teach' mode or through a real-time path control via pre-computation in an off-line Teach-Assist-Computer. The system would be particularly useful in small batch manufacture allowing the introduction of automation which would normally be uneconomical.

The research programme on application of an ASEA 60 industrial robot as an open-die forger aimed to develop an automated forging system (see Fig. 1) which takes full benefit from the inherent advantages of both the free forging process and the present generation of industrial robots.

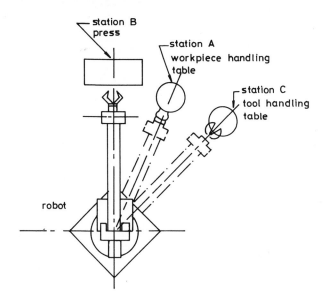

Fig. 1. *Plan View of the Equipment Layout*

A special gripper, consisting of wrist and finger mechanisms, was designed and constructed. One feature of the gripper was its ability to absorb the movement of the workpiece material, namely elongation, towards the robot arm during the forging operation. This was necessary firstly to avoid subjecting the robot to the considerable horizontal forces generated during the deformation and secondly to allow the return of the workpiece to its original position after the release of the press stroke, thus maintaining a retractable datum. The finger operating mechanism was mounted on a carriage which slid longitudinally on a pair of guide rails fixed to the base of the wrist mechanism. The diagram in Fig. 2 shows the gripper design.

Application of an ASEA robot

Initial assessment of the problem revealed that even in a simple six squeeze two pass forging the number of manipulative steps was quite large. However, the ASEA control system software is capable of accepting sufficient information, for far more complex forgings, using instructions such as repeat, pattern, jump, conditional jump and repeat patterns with datum shifts. This makes composition of quite compact programmes possible in order to perform numerous yet repetitive manipulations required in a forging operation. Each move is determined by the operator through

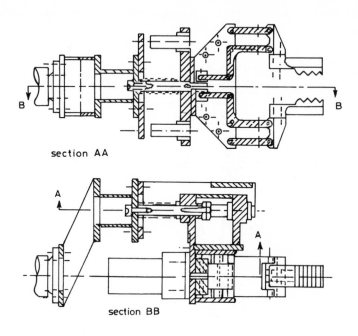

section AA

section BB

Fig. 2.   *Robot Gripper*

push button control on the portable programming unit and the position and speed is stored in the computer memory. After the programme is approved in automatic mode by a trial run, the operator usually commands the computer to store the programme on a magnetic tape so that it can be recalled when that particular forging job is to be performed. The use of special programming software allows compact programmes to be developed which need only about one third of the memory requirement, compared to the one composed by a point-to-point, step by step 'teach' method.

In a small batch production it is important that the time taken to generate a robot programme should not be prohibitively long or expensive. With these constraints in mind it has been suggested that in cases where an industrial robot interacts with a variable skill-intensive process, a system approach should be adopted in which the manipulation process is decomposed into small cells or modules of programme.

Some of these modules can be used repetitively without modification and others can be capable of selections and combinations of pre-recorded responses so as to provide choices of action in response to changes in process parameters. The computation of numerical values representing variables in each forging sequence may be performed on an off-line dedicated or non-dedicated mini-microcomputer or eventually by moving towards on-line, real time computer control.

Bar forging

Prediction of Forging Schedule
In order to enable the robot to perform a particular job in a pre-programmed manner, it is necessary as a first step to predict an optimum schedule for all press and manipulative actions in a proper sequence and with full details of spatial positions. The predicted schedule should clearly specify the optimum number of passes and squeeze ratio during each pass, indicating the tool gap setting between top and bottom dies, number of squeezes in each pass and the necessary manipulator movements.

Three fundamental conversions carried out in open-die forging with flat tools are (i) from a square bar of given length to a longer bar of a smaller square cross-section and (ii) from a round bar of a given length to a square bar of longer length, (iii) from a square or round section to a flat rectangular blade. In the present work, all these conversions have been studied.

The workpiece to be forged is usually much longer than the size of the tools and therefore, the operation consists of a series of squeezes applied from one end of the bar to the other with what is commonly termed a 'pass'. Several passes are usually required to complete the job, because, firstly both pairs of faces have to be worked as height as well as width has to be reduced, and secondly there are limits to the extent of reduction in one squeeze because of the danger of introducing defects.

Computation of the optimum forging schedule

The computation of the optimum forging schedule involved calculation of spread and elongation for each squeeze. The number of passes required to produce the finished cross-section was computed from the consideration of imposed technological restrictions and the number of squeezes for each pass was estimated from the concurrent knowledge of current length. A computer programme to derive optimum schedules was written in FORTRAN IV language and was run on an LSI. II microcomputer.

Experimental procedure

The experimental work during the study consisted of reducing the height and width of square bar specimens. Pure lead was used as a model material throughout to simulate hot steel. A number of different squeeze sequences were investigated as illustrated in Fig. 3, and it was shown that most consistent results were obtained by the forging schedule represented by sequence 2. The other three modes produced wavy unevenness or localised slipping at the free end and are therefore unsuitable for practical application. The general conclusion after this initial work was that robots should be suitable for open-die forging using open-loop control.

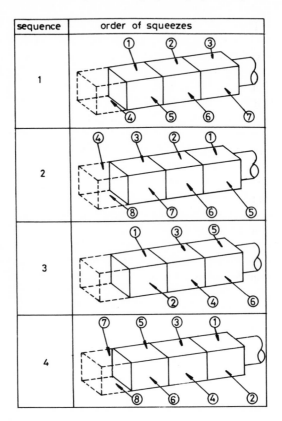

Fig. 3.   *The Different Forging Sequences Investigated*

The computer programmes mentioned above were used to predict the optimum forging procedures required for conversions in terms of reduction per pass, number of squeezes per pass and number of passes for the whole schedule. The press gap for each pass was manually fixed at the computed value for that pass. The robot was programmed by going through the whole sequence for this schedule.

Because the press was single acting there was need for the correction of centre-line movement of the workpiece during the squeezeing process. Moving the arm vertically by a distance equal to one half of the total reduction in a time equal to the period of the squeeze eliminated the possibility of undesirable side loading and displacement of the robot arm. As the press released the load at the end of the squeeze, the workpiece was lowered to its pre-squeeze level and could then be freely manipulated to the position for the next squeeze.

The robot was used to produce a number of similar forgings using the identical sequence of operations as programmed in the above manner. The height, width and length of the workpiece after each pass were measured and were compared with those predicted by the computer programme.

In industrial applications there is a high probability that both ends of the workpiece material will be required to be forged. For such eventualities a procedure was devised to remove the workpiece from the press and transfer it to a workpiece handling station. At this station the workpiece was released, reorientated, and then regripped at the previously forged end. The workpiece was then returned to the press and the initial forging schedule was repeated.

A separate programme to convert a short length of the reduced square forgings to 'crude' round cross-section was also taught to the robot by the manual 'teach' method. The sequence of conversion was as follows: (a) from the square to octagonal cross-section by squeezing at the diagonal of the square in four passes and (b) conversion of the octagonal cross-section to a 16-sided polygonal cross-section through rotational manipulation of the wrist between squeezes.

The dimensional consistency of the forgings was monitored and it was found to be good as was the smoothness of the surface. For comparison purposes, a few workpieces were forged by hand, the operator exercising extreme care and caution and the robot pieces, without doubt, were found to be of superior quality.

It was thus confirmed that it was possible to employ an industrial robot to open-die forge the basic shapes described in this paper in competitive time compatible with the heat retaining capacity of mild steel workpieces. The quality and consistency of such forgings are likely to be superior to conventional manual open-die forgings.

Ring forging

A common example of free forging is the manufacture of rings; forged ring blanks are the starting form for a large number of engineering components such as gears wheels, rollers etc. Conventional ring forging using a hammer or a press requires four to five highly skilled workers operating the equipment and manipulating the workpieces between and during forging operations. Application of industrial robots in ring forging could lead to a substantial reduction in the need for skilled manpower. This would undoubtedly be coupled with significant improvements in product consistency and an increase in productivity.

As a logical extention to the initial work on bar forging an attempt has been made to 'teach' the robot to perform ring forging. It was apparent that workpiece handling and manipulations in ring forging are far more complex than in bar forging. Ring forging involves imparting deformation to all faces of the workpiece accompanied by considerable changes in the workpiece shape from billet to ring form.

An important extention of the earlier work relates to the gripper design. For ring forging and most other free forging processes there is a continuous and progressive change in the workpiece shape during a compression cycle. For bar forging the gripped portion of the workpiece does not deform significantly but is displaced by the elongation of the bar. Thus in early work the gripper only required axial compliance. However, in billet forging it is generally desirable for the gripper to maintain a hold upon the workpiece, and so 'finger compliance' is required, i.e. the ability to allow the fingers to be displaced by the spread of the workpiece without damage and without loss of location. In addition the gripper must be able to accomodate large variations in workpiece size.

Ring forging operating sequence

The forging of a ring from a heated square billet is usually accomplished in two stages. In the first stage the billet is forged into a regular circular blank and a hole is pierced. Two alternatives are available for the second stage. A second stage using a forging hammer or press involves saddle forging the thick walled blank to a large diameter.

An alternative method of increasing ring diameter at the expense of wall thickness is to use the ring rolling process. This work is primarily concerned with the first stage of ring forging.

In order to understand the robot manipulation requirements it is necessary to break the forging sequence into elements (Fig. 4).

The breakdown of the ring forging process into elemental operations clearly illustrates the complex skills demanded from operators. Apart from controlling the press and selecting the proper reduction levels the operator is also required to perform complex manipulations of the workpiece on consecutive strokes so that the workpiece is presented to the dies with the correct orientation. The operators are normally equipped with a number of auxilary tools such as tongs and shackles to grip, twist, turn or topple the workpiece, most of the manipulations being carried out within the die space on the press. It should also be noted, that in practice some operations such as hole punching or saddle forging need a change of tooling and skillful tool positioning.

Human operators are endowed with adaptability, judgement, skill and flexibility in the hands and wrists to perform workpiece and tool manipulation. In attempting to replace human operators by an industrial robot, it is soon realised that the robot is devoid of many of the valuable human attributes, namely, adaptability, judgement, sensory feedback, and the extent of flexibility in hand and wrist motions. Even with the five degrees of freedom available in the arm and wrist motions of the ASEA IRb-60 it was aparent that many of the workpiece manipulations could not be performed within the press gap. Thus it was necessary to transfer the workpiece outside the press gap where the robot arm and wrist motions could be undertaken unhindered by the structure of the press. It was noted that time elements would be involved in such transfers and manipulations and that they should have to be taken into account in the overall planning strategy.

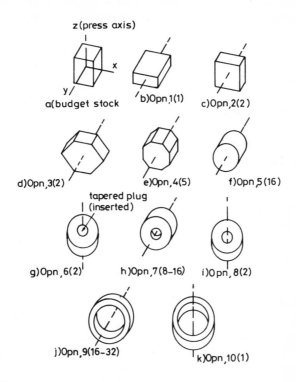

z(press axis)

a(budget stock   b)Opn,1(1)   c)Opn,2(2)

d)Opn,3(2)   e)Opn,4(5)   f)Opn,5(16)

tapered plug
(inserted)

g)Opn,6(2)   h)Opn,7(8–16)   i)Opn,8(2)

j)Opn,9(16–32)   k)Opn,10(1)

Fig. 4.   *Sequence of ring forging operations (Numbers in parenthesis indicates number of squeezes)*

Experimental forging and piercing a blank for ring forging

The principles of robotisation of free forging a ring in the manner described previously was studied experimentally by forging 70 mm × 56 mm sq. lead billets into 85 mm diameter by 40 mm thickness blanks with a 25 mm diameter hole pierced at its centre.

Programming the robot microprocessor on the modular basis required only 92 positional instructions and approximately 250 other instructions such as those needed for gripper operations, press operations, testing of flags and linking various parts of the programme together. The whole programme is estimated to have taken up around 2500 memory words which was nearly a quarter of what would be required on point-to-point programming. However, an important advantage to note is that the modular technique allows one to programme forging a ring blank from a range of billet sizes. Only the modules with variable movement lengths need to be modified in accordance with the changing dimensions, while the fixed modules can be left unaltered. In the next phase of the research programme, the resident micro-processor is proposed to be linked with a micro-computer which could be utilised for pre-computing of the variable movement lengths and for incorporating the necessary modifications in positional instructions stores in the robot memory.

Conclusions

1. The ASEA robot is capable of the complex manipulations required for the forging of bar products and rings using a schedule not unlike that currently performed by human operators.
2. The gripper design described in the paper was found adequate for the experi-mental programme, the designed compliance mechanisms behaving in the required manner.
3. It was found necessary to break the forging procedure into elements and to programme the robot using a series of modules. The building of the modules into a comprehensive programme appeared to be an efficient means of building long complex programmes.

Acknowledgements

The authors are indebted to ASEA (UK) Ltd., for their kind assistance and the loan of the ASEA 60 kg robot used for this work. Thanks are also due to staff at Doncasters (Sheffield) Ltd., particularly Mr. S. Nichols and Mr. B. Biddle, for their practical advice on forging procedures and their continuing encouragement of the work. The SRC must also be recognised for their general support of the Robot Forging Research.